ATOMIC ZOMBIE'S BICYCLE BUILDER'S BONANZA

BRAD GRAHAM
KATHY McGOWAN

McGraw-Hill

New York   Chicago   San Francisco   Lisbon   London   Madrid
Mexico City   Milan   New Delhi   San Juan   Seoul
Singapore   Sydney   Toronto

The McGraw·Hill Companies

Library of Congress Cataloging-in-Publication Data

Graham, Brad.
    Atomic Zombie's bicycle builder's bonanza / Brad Graham, Kathy McGowan.
      p.   cm.
    Includes index.
    ISBN 0-07-142267-6 (alk. paper)
    1. Bicycles—Design and construction. I. McGowan, Kathy. II. Title.

TL410.G68   2004
629.227 2—dc22   2003060436

8 9 0 DOC/DOC 0 9

ISBN 978-0-07-142267-3
MHID: 0-07-142267-6

The sponsoring editor for this book was Judy Bass and the production supervisor was Sherri Souffrance. It was set in Century Schoolbook by Ampersand Graphics, Ltd. The art director for the cover was Anthony Landi.

Printed and bound by RR Donnelley.

McGraw-Hill books are available at special quantity discounts to use as premiums and sales promotions, or for use in corporate training programs. For more information, please write to the Director of Special Sales, McGraw-Hill Professional Publishing, Two Penn Plaza, New York, NY 10121-2298. Or contact your local bookstore.

CONTENTS

PREFACE

There are many reasons why I wrote this book, but the most important reason is that it's necessary. After years of researching the Internet, the library, and bookstores trying to locate books on radical bike building to see what others have done, I realized none existed. I only found books on bicycle repair and maintenance, bicycle history, and bicycle racing, but not a single book that showed how one could alter a typical bicycle to make something new or extreme.

Although these books offer a large amount of highly accurate and technical information, none of them would appeal to the type of person who wants to dig right in and create something totally off the wall or amazing. The type of bicycle people I associate with don't care about proper pedaling cadence or the rolling resistance of a carbon fiber rim. They just want to know where to aim the hacksaw to make frame hacking quicker and more effective. We build things out of junk, we follow through on ideas that the "experts" say will fail, and we have a lot of fun doing it.

What type of person are you? Are you happy cruising around on a standard mountain bike, or do you want to roll down the street on something so wild that everyone you pass has to have a second look? Well, since you are holding this book in front of you, I will assume that you are the second type.

Another reason for this book is proof. I really wanted to show the world that this hobby is within everyone's grasp. I have no special tools or formal training, and I do not spend a lot of money (if any) on most of my projects, but look what I have built. Do I have some special gift or magic trick? No. Anyone can do what I have done and more with only a little patience and determination.

For this book, I built all of the projects in an unheated, uninsulated 10′ × 18′ garage during the winter months using only a welder, a grinder, and basic tools. Ninety percent of the raw materials were scrap bikes salvaged from the scrap yard or donated by friends. The other 10 percent was inexpensive electrical conduit bought at a local electrical supply store. Some bikes, such as the Marauder, were built from junk parts, and have carried me for hundreds of trouble-free kilometers at speeds that some riders can only dream of reaching on their expensive upright bikes.

So, there you have it. If a 34-year-old computer tech can build a custom bike in an unheated garage in the dead of winter from a pile of rusty old junk, then so can you!

Another reason I wrote this book is to spark your imagination and provide you with the basic ideas and skills to bring your ideas to life. You may want to build one or more of the projects presented in this book, and that would be a great way to get started, especially if this is your first jump into this hobby. But once you start to master the art of joining things together, there are no limits to what you can make from the bicycle scrap pile.

I have been creating wild and crazy bikes for almost 20 years now, and this book represents only a fraction of all the projects I have done over that time. Although my welding is a lot stronger and visually appealing today, I still managed to make a lot of working machines in my early days. When I was 14, I welded six different bikes together to form a horribly twisted bicycle built for six. My friends and I would stay out at night and race around town until it broke (usually in half), then drag it back to the shop and add more metal to it where it failed so we could ride it some more. We had a blast riding these crazy bikes, and entertained not only ourselves, but also the people who happened to see us ride by.

Now, 20 years later, I still have just as much fun on my crazy contraptions as I did when I was 13, and my designs have matured to the point that a catastrophic welding failure is no longer a worry at high speed. I can't see this passion to create custom bikes ever dying, and when I'm an old man, I will still be doing this, although my designs will probably include a lot more three-wheeled vehicles and fewer low-racer types.

I hope my book inspires you to turn your ideas into working projects. Don't let the complexity of any finished machine overwhelm you. Just remember that it is only made up of smaller parts, parts you can build. When something goes wrong or fails, don't give up. Keep trying until it works out the way you want it to.

Anyone with money to burn can acquire an exotic and expensive bike, but only determination, patience, and hard work will turn your ideas into reality. Bicycle building is a great hobby for all ages and skill levels. So, get motivated, and start designing!

ACKNOWLEDGMENTS

Writing this book was a team effort consisting of me and my significant other, Kathy McGowan (see Figure 1). I endured –30 °C temperatures for hours in an unheated garage trying to grind something useful out of piles of twisted junk in a matter of months, but I believe that she had the hardest part of this book project—keeping me on track and organizing it all. Without her support and organization skills, this book may never have made it to your hands.

Many of the projects in this book were refined with help from my son, the ultracool and fearless Devon Graham (aka "Roadkill"). How do I know if a frame will be strong enough? I give it to Devon for five minutes, and he will try crash it every way possible. Devon will test ride anything I make without any hesitation. If I make a bike that he thinks is too scary to ride, I won't bother taking the driver's seat.

Figure 1 Kathy McGowan (left) and Brad Graham (right) first teamed up as toddlers in 1971.

I really appreciate the support and interest of many Thunder Bay businesses, neighbors, strangers, friends, and family members. Special thanks to Adam, Brittany, and Christina Zuback for posing for pictures and helping me test some of my creations. Thanks to Rob Zuback and Nick Perna for being brave enough to ride the SkyCycle II. Jonathon Wilson, a reporter with Thunder Bay Television News, has profiled me and my projects on our local news station. He was also brave enough to ride the SkyCycle II and broadcast his experience on the air.

Thank you to Garry Felbel, owner of Garry's Automotive in "friendly Westfort Village," for letting my friends and me use his property after hours to test my bike projects. Of course without the amazing tolerance of my parents, Tom and Lillian, none of the projects in this book would have been completely finished. For months, they put up with a basement full of scrap parts and paint fumes, as well as many hours of hammering and clanging metal. They complained about the horrible mess I made, but they never actually made me stop. So, thanks for being cool. They get the Most Tolerant Parents Award.

Troy Way has also helped me enjoy my bicycling experience. He's always ready to make a wild machine, or scour the scrap pile for treasures. Troy also built a fast low-racer, and it's the only other nonmotorized vehicle on the road that gives me a competitive race. Troy's creations can be seen at www.bluepdoo.com.

Although not involved in the whole crazy bike movement, Steve Becotte needs to be mentioned for his ongoing support and friendship. We haven't become rock stars yet, but I know we'll end up where we want to be some day, and dude, I'm not going without you. Sorry about the horrific bike experience I put you through when we were young. That riverbank was much steeper than I thought.

Thanks also to Judy Bass, Senior Acquisitions Editor at McGraw-Hill, who encouraged me to write this book. I was never really interested in putting my projects in a book format, but I've been having a great time, and am already thinking of ideas for the next one. Thanks for believing in this project. Also, thanks to the great folks at The Cyclepath and Kinecor for their help, support, and interest in my whacky projects.

Lastly, I would like to thank all the people who tossed their old and unwanted bicycles into the giant scrap metal pile at the city landfill site. If you could only see your bike now! Anyway, keep them coming. I'm getting low on frames again.

BASIC TOOLS AND SKILLS

So, you have some wild and crazy plans to build your own bicycle and enough raw materials to start your own scrap yard, but where are you going to put it all together? You need a workshop, of course, and the better you set it up, the more efficient you will be. Even if your workshop consists of nothing more than the 8′ by 8′ tin shed in your backyard, it still needs to be set up properly.

Workshop Features

The first concern is power. You will be using standard plug-in tools such as drills, grinders, and saws, so having a standard 120-volt outlet is necessary. If your workspace doesn't have power, then get a good heavy extension cord and run power from the house. Usually, an exterior outlet will be fine and, normally, they are on their own circuit, but check it first. If you are running a large grinder on the same outlet that other appliances share, you will constantly trip circuit breakers or blow fuses. Also, household extension cords are not designed for outdoor use, so don't plan on burying them or nailing them along a fence with cable straps.

With power comes light. Trying to make a precision cut in a dark shop with only a 30-watt desk lamp is futile. Good overhead lighting is a must. Usually, for each 8' by 8' area, an overhead 150-watt light will do the trick. Even if you had to run an extension cord to your workspace, installing a permanent lighting system will not be a problem. You could install a few simple bulb sockets and run a plug from them into a power bar. This will be disconnected each time you remove the extension cord from the house. There are also premade bulb sockets with cords already attached, and these can be hung from the ceiling. A 150-watt bulb generates some intense heat, so install it properly, to avoid heating any nearby flammable objects or the building itself.

THE WORKBENCH

Now that you have power and light, you must plan out your work area. The most important place will be the workbench.

A good workbench will have a solid top made of either heavy wood or a metal plate and should provide plenty of room for your project. A 2' or 3' deep bench is good, but anything less than 2' may be a little too small, and even something as small as a bike frame would be hanging over the edge. If there is no bench in your workshop, one can be made very easily on a budget from 2' × 4's and plywood.

A plan for a simple, yet strong workbench is shown in Figure 1-1. I have not presented a rigid set of measurements because the size of the top and height preference determine the length of all the boards. This bench can be made in just about any size or height combination.

The first thing you need to do is figure out how tall to make your bench. This is usually a personal preference, but there are limits. If your bench is too short, you will be hunched over your work, and working with a vice will be difficult. If your bench is too tall, you will have to stand on a crate to reach the back, and this will be dangerous. A standard workbench height is about 3' from the floor to the top, so if you have no idea how tall to make yours, then just use the standard.

Second, how big do you want your bench top? Any size works. I made mine 12' long, and it fills one entire wall of the garage, but you may

Figure 1-1   A simple workbench plan.

not have that much space, so deciding on the length will be the most critical part before you start cutting boards.

The first step to making this bench is to build the top frame (see Figure 1-2). The front and rear boards are cut to exactly the length of the final bench, and the two end boards are cut to the final width you want, minus three inches. If you did not subtract the three inches, the bench would be three inches wider than you planned since this compensates for the thickness of the front and rear boards. Cut and nail your top frame so it looks like the one shown in Figure 1-2, with the end boards nailed so they are inside the front and rear boards. All boards here are 2′ × 4′s.

Now cut four boards to the height that you choose for the final bench. These are nailed flat against the end boards in each corner, as shown in Figure 1-3. Notice that the top of each leg is flush with the top of the frame. At this stage the bench can stand on its four legs, but there is very little strength, and the unit is very wobbly. In the end, the bench will be either nailed against the wall, or cross braces will be added to give it strength.

Now you're left with the bottom four boards. You will need to cut two lengths that are the same as the top frame ends, and two that are six inches shorter than the top frame, front and rear. These are added to

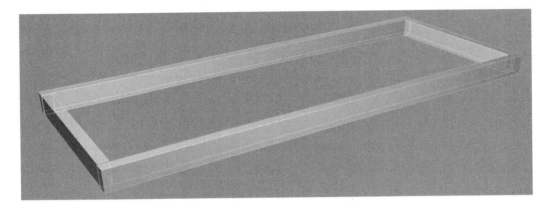

Figure 1-2 Build the top frame to the size you want.

the bottom (see Figure 1-4). Nail the two longer boards in between each front and rear leg set, then nail the shorter boards against the two side leg sets.

At this stage, you can install your bench into its final destination. You will need to nail it to the wall, so it cannot move from side to side. It's not yet strong enough to stand on its own. If you cannot nail

Figure 1-3 Bench legs added to the top frame.

Figure 1-4 Completed bench frame.

your bench to the wall, or it will be in the middle of a room, then cross bracing must be added. One brace running from the top rear corner, diagonally across to the other rear bottom corner, and one diagonal brace on each end of the bench will be enough to give it that added strength.

Your bench now needs a top. 2' × 4's or 2' × 6's work great for the bench top and can take a lot of abuse from hammering and grinder sparks. This is also a good material on which to bolt down a vice or bench-top grinder. If you want to use a solid board for the bench top, then use at least a ¾ inch thick plywood, not particle board, Aspenite, or some other composite board. Your bench top needs to be very solid because you will be using the hammer a lot to remove rust from parts, or to bend fabricated pieces. Nail your boards lengthwise along the top of the bench, as shown in Figure 1-5. If you are using 2' × 4's or 2' × 6's, you may want to leave a small gap between the boards so dust and grinder shavings can fall through, making it easier to clean up (see Figure 1-5).

Shelves can easily be added to the bench by adding another 2' × 4' across the middle of each set of end legs, then nailing plywood or more 2' × 4's across them. If you made your bench longer than four feet, then you may also have to run 2' × 4's along the length for ex-

Figure 1-5 Top added to the workbench.

tra support of the shelf, similar to the way the top frame is constructed.

Now that your bench is complete, bolting a large vice to one of the ends is a good idea. When you need to cut or grind steel, you need some way to hold it down securely, and this is the easiest method. Do not use nails to hold down a vice, as these will work loose after time. The best method for securing a vice is to use large bolts and lock washers. If you used plywood for your bench top rather than 2' × 4's or 2' × 6's, then you may want to add a small 2' × 4' plate under the plywood to bolt the vice to for extra strength.

Now, if you can find a convenient place for your toolbox and other tools, you will have created a work environment that will lend itself to comfortable and safe working conditions. All you need now is a couch and television, and you can live in the garage (something the hardcore builder likes to do)!

MANUAL AND POWER TOOLS EVERY BUILDER NEEDS

There are certain tools that every builder must have, whether you plan to just do the odd bit of tinkering or go into mass production of your latest creation. As I built the projects for this book, I did not use any tool that is not included in this list. None of these tools require any special skills to operate, nor do they cost a fortune.

- Vice. A good vice to bolt down to your workbench.

- Basic socket and wrench set. You can get these from any department store and, chances are, you already have these, so I won't ramble on about them.

- Adjustable wrenches. Handy for those quick jobs or to hold a bolt while you turn a nut with a socket. You get the idea. Having an adjustable wrench large enough for removing a bicycle fork nut is a good idea. A wrench this size will need to open up to at least 1¼", but something that will open to 1½" or 2" would be ideal. A few smaller ones are always handy as well.

- Large hammer. Yes, that cute little hammer you use to whack nails into wood will probably get you by, but a real iron-worker's hammer is what you need to bend a plate of steel or get that rusty old crank arm free. The hammer you want will be about twice as heavy as a typical claw hammer; one end will have a flat head and the other will have a ball head.

- Large side cutters for cutting old spokes and rusty cables.

- A tape measure and steel square. You want things to be straight, right?

- Angle grinder. This tool is a must. A six or eight inch hand-held grinder is perfect for cutting tubing and cleaning welds. This may be your most-used tool if you plan to do a lot of metalworking. Purchase several grinding disks and a few cutting disks (also known as "zip" disks) as well. A good set of safety glasses or full-face shield will also be necessary.

- Hand drill. Try not to get a cheaper one. Discount plug-in drills from hardware stores have barely enough power to drill a quarter inch hole before they need to cool down. Do some research and find a good brand name. A drill that can take a 1/2-inch bit will come in handy.

Besides the welder (described in Chapter 3), these are all the tools you will need to create all the projects in this book and many more. Of course, if you have access to better tools such as a drill press, chop saw, or even a lathe, then you will have an easier time making your designs come to life. Remember, I built all these projects with only the above tools in a small unheated garage in mid-winter, so don't panic if you don't have access to a 2,000-square-foot heated-floor garage with a washroom!

PROPER MEASURING AND CUTTING TECHNIQUES

If your raw materials aren't straight, your final project won't be straight; it's as simple as that. Measuring a length of pipe and cutting it off at a nice 90 degree angle is not hard to do if you know a few simple tricks.

First, don't try to copy the pros. You know the type—the guy who throws a pipe in the vice, grabs a hacksaw with half the teeth missing from the blade, cuts through the thing like butter with one hand while he drinks coffee with the other, and still manages to get a perfect cut. Although this can be done with enough practice, let's be realistic.

Cutting a square tube with a hacksaw or grinder disc is not an art—it is done by marking all four sides of the tube, then cutting each side on the line. If you just try to cut right through the tube from one starting point, you will usually end up cutting on an angle. You cannot see where the blade is going at the back of the cut, so getting it straight will take a bit of luck. Here is the best way to get a clean cut through a square tube. Measure the length to be cut and make a mark on one side of the tube. Take a square and draw a line across this mark from one side to the other, then transfer this line to the next side of the tube with your square, and so on until you are back at the original line (see Figure 1-6).

Place the tube in the vice and only cut the side of the tube facing upward right on the line. When this is done, turn the tube and cut the other sides until the length is cut. Even though you may have made a small mistake on a few of the cuts, the overall job is still fairly straight since you cut each side from a fresh line. With a little touch-up on the grinder, the end will be as square as if it was cut in a chop saw. Try cutting through a tube without turning it in the vice and compare the results; chances are good that the first method will always be the best.

Cutting a round tube is can be done in a similar fashion, although you won't need a square. Just mark a line where you want the cut, then wrap a piece of duct tape around the pipe so the edge meets the line to be cut, as shown in Figure 1-7. Since the end of the tape has to meet the start, it makes a perfect line around the tube. When you are cutting the tube, again, just cut the top a little way in, turning the tube around in the vice so you can always see the line to follow.

You may want to purchase a tube cutter for small round tubing (up to 1½). These are inexpensive and make perfect cuts with very little effort. Place the tube cutter on the tube and twist it around until the little blade cuts the tube. This little tool will make a perfect cut every

Figure 1-6 Marking a square tube for cutting.

time, but it only works on perfectly round tubes, not square ones or the oval tubes found on some heavier bike frames.

When cutting several pipes to the same length, always measure each line separately. Avoid tracing a line from another cut piece, because you will transfer any mistake along to the next piece and so on until the mistake is very large. Also, each time you do this, you add the thickness of the marker or pencil into the measurement and, after several cuts, this can add up to quite a bit.

SAFETY MEASURES

It's important that you have proper safety equipment before you start any project. When using a power tool such as a grinder, make sure that you use extreme care and diligence. This section describes some recommended equipment and safety procedures while working on projects.

Grinder

A hand-held grinder can be dangerous. In fact, out of all the tools you will use for doing this type of work, it is the most dangerous of them all.

Figure 1-7 Marking a round tube using tape.

Eye protection is an absolute must when you're using a grinder. There are thousands of red hot, sharp microscopic filings flying in every direction as you cut and grind. A full-face shield is the best choice to deflect this bombardment because it will cover your entire face. If you plan to use safety glasses instead, find a pair that has some side protection, since these little shavings are flying in all directions. Sometimes the filing will bounce off the bench or the floor and fly back up under your safety glasses. This is why a shield is better.

Positioning your face during grinding is also something to think about. If a grinder disc decides to fly apart at 10,000 rpm, it will smash your face shield and safety glasses, so try not to put your face directly in line with the disc, although sometimes this cannot be avoided.

Ear protection is also necessary. After continuous use, the intense sound levels produced by a grinder will damage your hearing, so a good set of ear protectors are needed. Those little plugs that cram into your ears do work well, but they are not made to be used over and over, so get a real set of ear protectors. They will last for years.

Good gloves should be worn at all times, and avoid any loose clothing that can get wound up in the disc. If you have sleeves, make sure they are not flapping around or loose. Always hold on tight to the grinder! I have had a grinder ripped from my hands a few times as it snagged, and it is a scary event, since you don't know where it is going when it is on a holiday from your grip. I was lucky, since the grinder ended up bouncing along the garage floor, not into my leg, but now I always take a firm grip with both hands.

It is a common practice to remove the grinder's safety shield because that way you can use discs that normally would not fit onto the grinder. This is a bad idea because the grinder's motor will be under stress, and the discs will be placed too close to your hands. A 6-inch grinder may accept an 8-inch disc with the safety shield missing but, trust me here, don't do it.

Drill

Compared to the grinder, the drill may seem tame, but there are still a few things to watch out for. Although red hot shrapnel will not be flying in all directions, wear your safety glasses to protect your eyes from other debris while drilling. The biggest danger when using a hand drill is not the drill itself, but the work you are drilling.

If a freshly cut piece of steel comes loose from the vice, it will spin around with the drill bit, and this could cause injuries if it spins into your hand or face. This is especially true with a quality drill because it has more than enough power to break your wrist, so be careful and make sure that everything is secure.

Hacksaw

Keep your hand out of the stroke path and wear good safety gloves. When starting a new cut, the blade sometimes slips from the work and the blade always seems to find its way to your other hand. So, make sure that you keep your free hand out of the way if you like your knuckles without scars!

Hammer

Never use a hammer with a loose head, and always wear safety glasses when smashing against steel. The steel on bicycle pedals is sometimes hardened like the head of the hammer, so small bits of steel may fly around.

Conduit and Galvanized Steel

Some conduit and steel is galvanized. There is a thin layer of plating that will burn off during welding and torch cutting, producing poisonous zinc oxide fumes. Inhalation of these fumes may cause short-term physiological problems similar to a flu, which can be easily avoided by using proper ventilation.

Ask your supplier if the tubing you are using is galvanized, and if so avoid welding indoors, or make sure you blow the fumes away from your face. I always make small welds on galvanized tubing, usually holding my breath if I am very close to the work. Once the thick white smoke has disappeared, the fumes are gone. Welding outdoors would be the best method to avoid the fumes.

Things That Make Sparks

Welders, grinders, and cutting torches make small hot sparks that can smolder for many hours before a fire starts. It's always a good idea to hang around the shop for an hour or two after you have been welding or grinding, just to be safe.

If you work in your garage, keep all gas tanks and oily rags safely away from the path of grinder fodder, as these things will light up very easily. Grinder sparks also burn through fabrics at a great distance. If you aim the sparks from your grinder at a lawn chair 50 feet away, there will be hundreds of small holes in the fabric within 10 seconds, so watch where you send your sparks. Grinder sparks can also light clothes on fire. I now own 20 sweatshirts with the sleeves and front burnt out of them, and this is something you'll have to be very aware of unless you can find some fireproof clothes.

Paint Fumes

Paint fumes come in two flavors: "freshly squeezed" and "charbroiled." Freshly squeezed paint fumes are the fumes that come right from the can as you paint your latest creation. These fumes are heavy and can slowly make you sick. If you are painting indoors, you will need a good ventilator with changeable filters. A real ventilator fits snuggly over your mouth and nose, and sucks air through two filters on each side. Go to a real paint store and ask for one. It is worth the extra cost.

When painting in an area with less ventilation, open as many windows as you can, paint in several steps, and stay clear of open sparks or sources of ignition, as heavy paint fumes can cause an explosion.

The charbroiled flavor of paint fumes is what you smell when you torch or weld painted or bicycle frames with stickers on them. These fumes are highly toxic, even in small quantities.

Try to clean the painted area with a sanding disk or your grinder first, to cut back on the fumes. The area needs to be clean at a distance of about 4 inches from a weld in order to stop the paint from burning. If you have to weld painted steel, hold your breath, make small welds, then head for some clean air to catch your breath. Do not breathe in burning paint fumes; they are far worse than the freshly squeezed type.

If you are painting or working with toxic chemicals, it's time to take a break if you get that "funky" feeling. Prepare the area so that it has better ventilation, then try again.

I hope you have a fun and safe time in your workshop. Use common sense when it comes to power tools. If it seems too loud, it is too loud. Wear safety glasses when necessary, and never trust your power tools or they will bite you when you let your guard down.

If a piece of metal is flying through the air, be careful. Never use dangerous tools like a lathe or oxy-acetylene setup without proper training. Your first error could be your last.

2

ACQUIRING RAW MATERIALS

Now you have a shop, tools, and a plan, but what are you going to build your bicycle out of? You could head down to the local bike dealer and purchase all new components and tubing, but unless you have a lot of money, this seems highly unlikely.

Sometimes, your design may not even be rideable in the end, so it would not be wise to put on a set of $3,000 carbon fiber rims and a $1,500 crank set before you even know if you are going to make it to the end of your driveway.

Even my high-speed "Marauder" project was made from junk at the beginning, and most test rides were done with cheap rusty components and a set of slightly warped wheels before I started upgrading to the final product. When you make a functional bike out of junk, just imagine what it will be like when you put on some decent components.

Anyone who has ever played with Lego blocks knows that your raw materials pile can never be large enough. In my garage, 80 percent of the space is taken up by the pile of scrap bikes, leaving only a small place to work. Having a large pile of different bits and pieces sure makes the design stage a lot easier, especially if you do most of your designing by trying rather than by planning on paper. If your only goal is to produce one really cool bike, then, sure, you could get away with just buying what you need. But for most builders, this is a hobby that will never end, and you will be churning out more radical creations than you even have space to store.

Landfill Sites

Let's get right to the point. You need to locate the landfill sites ("dumps"), scrap yards, and back lanes in your community. Most city or rural dumps have a specific area set aside for metal and recyclables. This is where you will find a lot of bikes and good metal parts. On a warm summer weekend, this metal scrap pile can grow into a large mountain and contain many bikes, some in fully working order. In Figure 2-1, Troy Way pulls out a bike frame from the pile of scrap at our local dump.

Sometimes, the dump may have rules about scrounging for parts. If this is the case, you won't be able to spend two hours taking the bent rims off the bikes, and you will have to take what you find as is.

Learning to shop at the dump is an art. You must know when to go and who to avoid. Most of the people who work at the dump don't mind if you take some junk with you, as long as you don't get in the way of regular traffic or try climbing up the scrap pile to dig something out of the top. This can be dangerous, and some dumps have strict rules because or liability issues. There are also certain times when the pile of scrap must be bulldozed away and, after this is done, most bicycle parts will be bent into a twisted, unusable mess. If you can figure out when the bulldozer makes its run, you can get to the good stuff before it's crushed into useless junk.

If you are told that scrounging isn't normally allowed, then leave without a fuss, go to the dump's main office, and ask for permission. Sometimes you will be allowed to go ahead; they may even let you pick through their own pile of "good junk."

There are also many "fringe" area and camp dumps. Although they may be a long drive away, sometimes there can be just as many good finds there as at the busier city dump.

Figure 2-1   A city dump is a gold mine for unwanted bicycles.

Your Neighborhood

If you don't have a vehicle to haul things from the dump, there may be a closer source for parts that you can get to on foot, or by bike—back lanes. In many communities, especially those with older subdivisions, there are back lanes. When I was young, I would wind my way up and down the back lanes looking for bike parts, and usually found one or two unwanted bikes on a typical run. I had a crude trailer made from other bike parts and a washtub in which I would tie down all the junk I found to haul home. After a month or two of hunting, my scrap pile became quite large.

Donations

The last good source of free bikes that I know of is donations from friends and family. If you know people who are avid "junk" collectors, they may be happy to donate anything to you just to see what it turns into. After all, your hobby can be someone else's source of entertainment! So put out the word to everyone you know that if they've acquired something made of metal, don't throw it out.

Yard Sales, Auctions, Pawn Shops, and Stores

So, you have searched through the dump and scrap yards, but you still need more materials. Yard sales are a good source of inexpensive bike deals, especially if you get there right before it's almost over. The last thing a yard seller wants to do is haul an unwanted bike back into the basement, so now is your chance to turn the $20 price tag into $5!

Local auctions are a good source for a good bike deal as well. People will usually only bid on the newer, fancier bikes. These are the kinds of bikes to avoid for your scrap pile because you can't weld aluminum or carbon fiber. Once the auction ends, there may be a lot of older bikes unsold, especially at some auctions that local police stations hold, and you may end up leaving with a hundred or more bikes for next to nothing if they just want to get rid of them.

Pawn shops are not usually a good place to find bikes. Once in awhile, you may find a real gem of a bike loaded with all the best going for a good price, but depending on where you live, there may not be much of a selection. Check out the pawn shops in your community and check back regularly if they do have bikes for sale.

If you have a project that is worthy of a set of half-decent rims, new brakes, and a crank set, you may actually want to have a look at a department store bicycle isle. Some department stores do carrying bikes with medium-quality rims, crank sets, and components. If you can get an entire bike for $200 that has half-decent aluminum rims and a nice crank set, this may be a better deal rather than buying the same parts at a bike dealer, even if you throw away the rest, or add it to your scrap pile.

Bicycle Shops

Now that I have covered all the good placed to find scrap parts and low-end parts, sometimes there are situations when you need the best.

When the Marauder was complete, it was obvious that the typical caliper style brake would not be enough to stop such a fast vehicle, so I had to spend money on a hydraulic disc brake setup. Quality-made parts are not cheap, but if total performance is what you want, then be prepared to open your wallet.

Find a bike dealer that you feel comfortable with and who understands your needs. Some shop owners will not only give you great advice for your "unusual" project, they will also take a great interest in your ideas and might give you a deal on parts. Some bike shops are only interested in making a fast sale, and have no time to give advice or listen to your "oddball" needs, so avoid these. A good signal that a bike shop has an open-minded attitude will be the style of bikes on display.

I was lucky to find a bicycle shop with owners that appreciated my unusual cycle creations, and always took the time to help me find what I wanted (see Figure 2-2).

Figure 2-2 Find a local bike shop with knowledgable staff.

When it comes to selecting quality components, I always ask the staff at The Cyclepath in my community for their advice, and they enjoy seeing what strange contraptions I will come up with next. It's always a good idea to get expert advice when it comes to cutting-edge technologies such as carbon fiber wheels or disc brakes because these are things you will rarely find in your usual scrounging hot spots.

Bring your designs or drawings to the cycle shop, and find someone who will help you select the proper parts or give you new ideas you may have overlooked. A fresh idea can do wonders for a project.

When you do start building your parts pile, try to organize the parts into some order, by stripping as much off a bicycle as you can rather than just piling them up into a 6′ hill of chaos. When you need that "cool" set of cranks from the bottom of the pile, you will see what I mean. Your garage can hold a lot more bikes when the wheels, cranks, and frames are stacked in separate piles, so make it a habit to strip down any full bike before you put it away, this will save you time in the long run.

Electrical Conduit and Tubing

For some projects, you may want to build a custom frame, and hacking apart old bicycles for the tubing may not be adequate, especially if you need a lot of the same size tubing or a longer length than can be found in a typical frame. Projects like the Highlander use tubes that are a lot longer than the ones in an ordinary frame, and trying to weld several lengths together would be a tedious job.

The SkyCycle, for example, has over 30′ of 2-inch round tubing in the frame, and for this you would have to chop the main tubes out of more than 15 large mountain bike frames! When I first started building custom frames that needed longer tubes, I would scour scrap yards for thin-walled tubing, but this can be hard to find. Most scrap yards mainly deal in construction-grade steel that is very heavy.

Of course, you could special order a 20′ length of brand new 1/16-inch-walled chromoly tubing, but this would cost a lot.

Let me introduce you to the best source of inexpensive tubing you will ever find—thin-walled electrical conduit or "EMT." This is the tubing that carries electrical wiring around commercial buildings, and it is about the same thickness as standard steel tubing used in bicycle construction. This conduit comes in sizes from ¾ inch to 6 inch, and is measured by the outside diameter. Figure 2-3 shows some of the com-

Figure 2-3 Thin-walled electrical conduit (EMT) in various sizes.

mon sizes of EMT. Many of the projects in this book, such as the Marauder and Hammerhead, use the very common one-inch conduit.

Thin-walled conduit can be purchased in 10-foot lengths (standard) from any electrical supply store, and it will cost you anywhere from three to ten dollars a length depending on what thickness and quantity you need. EMT is made from mild steel and can be easily welded, although some of it comes galvanized, and proper ventilation is necessary when welding it. EMT can also be bent into very smooth curves with an inexpensive manual conduit bender also available from the electrical supply store. With a bender, very fancy frames can be made with little effort by using curved shapes in the construction. EMT can also be cut in a jiffy using a simple hand-held pipe-cutting tool.

Although thin walled conduit is an excellent source of simple frame building material, sometimes you may need a stronger tube for projects such as the Marauder, or a very light tube for a racing project, possibly aluminum.

For special tubing like this, you will need to shop at a steel supplier, and make sure you bring along your wallet, because thin-walled tubing is not cheap, especially exotic materials such as chromoly or aircraft-grade aluminum. When purchasing new tubing, make sure you know exactly what you want, because the counter person is not going to give you a "really strong tube for your low racer" if you ask for one!

For sheet metal, such as what is needed for the Highlander's fender, any store that deals with industrial heating should be able to set you up with all you need. These stores make heating ductwork out of this sheet metal, and will usually have small scrap pieces you can buy fairly cheaply, or get for free. Again, watch for galvanized steel, and if you are not sure, ask.

Small odds and ends can be found in scrap yards and at the dump, and a lot of welding shops will let you dig through their scrap buckets if you ask them. You can never have enough "steel tidbits," so if you plan on doing a lot of building and welding, keep a few tubs full of steel odds and ends handy if you have the space.

WELDING BASICS

In its most basic form, an electric welder is a device that melts two pieces of metal together using high amperage delivered at the end of an "electrode" (rod) into the work. Although the technology that makes this process happen is technical enough to fill a book twice this size, I will deliver it to you in simple terms, the same way I learned it.

If you have ever had the misfortune to see an electrical short, either from bad wiring or some type of electrical equipment failure, you will have remembered the zapping sound, puff of smoke, and burnt metal that usually follows. This is basically what a welder does, but in a controlled and expected manner.

Arc Welding

The welder places a ground clamp onto one of the pieces of metal to be joined, then feeds a conductive electrode (rod) into the area or gap between the two pieces to be welded. Because the ground clamp and electrode are the two ends of a complete circuit carrying voltage and high amperage, a dead short happens at the end of the electrode, creating intense heat, melting both the electrode and surrounding area into a single fused piece. This type of welding is called "arc welding." The rod is not actually touching the work (parts to be welded) during the welding process. It is held a small distance from the work so that a very hot arc (electrical spark) can form. This arc produces the heat needed to melt the metals together (see Figure 3-1).

If the rod were pushed directly onto the work, a "dead short" would occur, an arc wouldn't form, and there wouldn't be any heat to weld the

Figure 3-1   An arc is formed between the electrode and the work.

metals. This is a common problem when you are learning to arc weld; it is known as "sticking." Sticking occurs when the rod is pushed too hard into the joint to be welded, causing the rod to stick to the work.

An arc welding rod is a thin metal rod coated with a material called "flux." This material melts as the rod burns away, giving off a protective gas that protects the newly welded area from the effects of extreme heat and oxidization. The flux also helps keep the arc performing smoothly, cutting down on sticking. If the rods were not coated with flux, the newly welded area would be damaged by the effects of oxidization, and would be prone to failure or rapid deterioration.

Some welders do not use flux-coated rod but instead feed compressed gas into the weld area during the welding process. These types of welders are called "wire-feed welders," but we will be only focusing on the basic arc welder here because it is the most basic type of welder to operate and the most affordable for beginners.

Although welders fuse metals together using incredible amounts of amperage, it is almost impossible to be electrocuted by an arc welding machine. The reason is that inside a welder there is a large transformer, a device that steps down the dangerous voltage in exchange for higher "amperage."

Amperage, the strength of a current of electricity, can seriously harm or kill, but only if a certain voltage is also present. A car battery can emit more than a hundred amperes (amps), which is much more than can be delivered from a standard wall outlet, but you can touch both terminals and not feel any electricity. How is this possible? Simply, because there is not sufficient voltage to deliver the amperage to your body. Of course, don't drop a metal object across a charged car battery's terminals or you will have a smoke show you will never forget!

Using an arc welder requires a little practice and patience, but is far easier to learn than any other form of welding. The hardware can be purchased from most retail outlets for a few hundred dollars or less. Anyone can learn to make a clean and strong weld with only a few nights of practice

Types of Welders

There are a few different types of arc welders available, and it helps to understand some basic concepts before you buy.

An arc welder is easily identifiable among other welders because it looks like a box with a few knobs on the front, as shown in Figure 3-2.

Figure 3-2 A typical arc welder or stick welder.

There will be no compressed gas cylinders, no spools of wire on top, no foot pedals or any other fancy options needed to make it work.

An arc welder may also be called a "stick welder," "rod welder," or "buzz box," depending on whom you ask. When inquiring about an arc welder, you will usually be asked what level of "input voltage" and "amperage range" you want. Input voltage refers to the type of outlet you plan on plugging it into—either a standard 120 V wall outlet or a 240 V outlet (similar to your household dryer outlet). The amperage range refers to the welder's output power (in amps) and, for the type of metal you will be welding, 100 amps would be more than enough.

The 120 V (standard wall outlet) type of welders are the most inexpensive types, and will have enough power to weld almost any thickness off steel that would be used in bicycle construction. Another advantage of this type of welder, besides the cost, is its size. These welders are no bigger than a large microwave oven, and can be moved around by one person easily. The disadvantage of these welders is the output power. Although an arc welder is fine for bicycle frame building, it wouldn't be powerful enough for projects with larger steel components, such as boat trailer frames or motorcycles.

A 240 V welder is a commonly used arc welder, and it is quit a bit larger, with an amperage rating of at least 200 amps. This type of welder requires a special plug that is easy to install, but your building must have 240 V wiring. They range in size from 2' to 4' tall or more. These welders cannot be easily moved around unless they are on wheels, and they weigh 100 lbs or more. The advantage of this type of welder is, of course, power. At 250 amps, you could weld steel plate thick enough to build a ship's hull.

The 240 V welder is the best unit to buy, as long as you have a place to put it and the proper electrical outlet to plug it into. This type of welder will produce a smoother weld, and will have enough power for any job you may want to do. Table 3-1 compares the features of both welders.

Another question you may be asked when purchasing a new arc welder is the output type—AC (alternating current) or DC (direct current). AC is the type of current that comes from your wall outlet, whereas DC is the type of current that comes from batteries. Welders that output DC are usually more professional and also more expensive. A dedicated welder will usually prefer DC for most work because it can produce a smoother weld and use a larger assortment of different rods for specialty work. An AC welder is the most basic type of arc welder and, for the hobbyist, all that is usually required.

I have used both types of welders (AC and DC), but chose a basic AC 240 V type for my shop, and this has been just fine for building just about anything. Although a DC welder can produce a slightly nicer fi-

Table 3-1 Comparing 120 V and 240 V welder characteristics

	120 V Welder	240 V Welder
Input power	standard wall outlet	special welder plug
Output power	0–60 amps	0–200 amps
Portable	yes	no
Weld quality	medium	high
Cost	low	medium

nal weld, a moderately skilled welder will be able to produce a much better weld on a simple AC machine than an unskilled welder could produce on an AC, DC, or wire-feed welder.

If you are new to all of this welding terminology, make your purchase at a welding supply house. The equipment sold there is only marginally more expensive, but the quality of a brand-name welder, combined with the advice of a knowledgeable salesperson, is worth the extra cost. A welder is a tool that will last for a long time, so choose the one that is suitable for your purposes.

Other Welding Equipment

Besides a welder, you will need to purchase a few other small items before you can make the sparks fly. You will need some protective gear consisting of a welding helmet, welding gloves, and safety glasses, as shown in Figure 3-3.

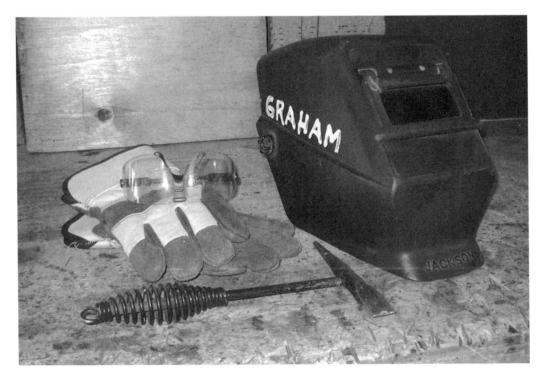

Figure 3-3 Welding helmet, gloves, and safety glasses.

Although some welders sold in larger retail stores may include a welding helmet, it's a good idea to get some advice on the type of lens you will need for your work. Lenses come in different shades and are denoted by a number. The higher the number, the darker the shade. All welding lenses protect your eyes from the damaging rays emitted from the welding arc, such as ultraviolet rays. As your power level setting determines the intensity of the arc, darker shades will be needed as intensity increases. Although it's always a good idea to talk to a knowledgeable salesperson to select the proper shade, Table 3-2 outlines recommended shade numbers based on amperage settings.

What does this mean to you? Well, nothing if you don't know much about setting the power level on your welder yet! To weld the light tubing that makes up most bicycle frames, amperage settings between 50 and 80 are commonly used, depending on your skill and brand of welder. Therefore, a lens shade of 10 would be fine. Shade darkness is not set in stone, and a beginner may want to choose a shade 9 rather than a 10, just because it will be easier to see the work. Of course, a shade 9 would be way too light if you had your amperage set at 500, and you would feel like you were staring at the sun!

Welding gloves are heavy, heat-retardant gloves that cover a large area of your forearm; this helps keep those hot sparks off your sleeves and out of the gloves. Although bicycle frame welding requires only requires a low power setting, there will still be sufficient heat produced to easily burn a standard pair of work gloves. The hot sparks flying from the welding rod can find their way into areas such as your sleeves, open collars, and pockets, so having gloves that at least cover your work shirt sleeves can help reduce those painful experiences.

A good pair of clear safety glasses or face shield will also be necessary. After a weld is completed, there is a thin layer of hardened flux from the rod coating the top of the weld area. This flux is removed with a "chipping" hammer, and this process sends hot flakes of flux in all directions, so you must protect your eyes. Some welding helmets allow

Table 3-2 Amperage settings and recommended shade numbers

Amperage Setting	Shade Number
20–40	9
40–80	10
80–175	11
175–300	12
300–500	13

Figure 3-4   A small box of 6013 3/32″ welding rod.

the dark shade lens to be lifted up, and have a clear glass shield behind the dark lens. This type of helmet will do the job of safety glasses during the chipping of a weld.

The last thing you will need is welding rods—a lot of them! If you are just starting out, you may want to buy a big box of rods, so you can practice making things stick together. It's cheaper to buy rods in bulk from a welding supply store rather than in the small pack you will find on a department store shelf.

Welding rods come in as many flavors as candy, and an entire chapter could be written on choosing the correct rod. To keep things simple, we will use "6013" rod because it is fairly generic and sold in stores that also carry welding equipment. 6013 is a basic and inexpensive rod used by beginners or for general work. Most of the tubing you will be welding will be about 1/16″ thick, so a 3/32″ 6013 welding rod will be fine. The outside of the box will have the rod type and size on a small label, as shown in Figure 3-4.

Basic Welding Skills

Welding is a skill learned by practice. No amount of theory or information will make you a good welder, only hands-on experience. I will present you with just the basic information needed to make a good weld, but you must put in the time and learn it yourself. If you want to become a professional, there are many good courses offered at colleges or even night schools. You will not only learn how to arc weld, but also how to use a cutting torch and all of the welder theory you will ever need.

STRIKING THE ARC

Before you start welding bicycle tubing together, it's a good idea to practice on thicker steel because the thinner the tubing, the harder it is to make a good weld without burning through. A few pieces of scrap 1/8″ plate or angle iron will make a good surface to test your welder with. Put a welding rod into the handle, set your welder's amperage dial to around 80 or so, then place the ground clamp onto the part you will be welding (see Figure 3-5).

The hardest part of learning to weld is getting an arc going without having the welding rod stick to the metal. Imagine that the rod is a stick match, and that you will be striking it along the work to light it. Hold the handle at a 45-degree angle with the electrode trailing your hand, as shown in Figure 3-6, then quickly strike the tip of the electrode along the metal using wrist action so an arc can form. If you push too hard, or your welder's amperage dial is set too low, the electrode tip will stick to the metal, and you will have to move it back and forth to free it. Welders call this a "false start."

If you had to move the electrode back and forth, the protective flux may have come off the end of the rod (see Figure 3-7), exposing the

Figure 3-5 Always place the ground clamp on the work.

Figure 3-6   Striking the arc.

Figure 3-7   Bare rod exposed due to chipped flux.

bare rod. It is very hard to get a bare rod to form a good arc, so you may want to cut the bare end off or get a new rod. Once you are more experienced, starting an arc with a bare rod will be possible but, for now, don't make things harder than they need to be.

If you managed to get an arc started without much difficulty, you will notice that there is a range of about one inch that you can move the rod away from the work and still maintain the arc. Try to keep the rod as close to the work as possible without pushing it onto the surface, as this may cause a stick. With the rod held at a 45-degree angle, and a good, close arc going, drag the tip along the metal for about five seconds, traveling about one inch in a straight line. If all went well, you will have a "bead" of weld that looks like the one in Figure 3-8.

Don't forget to chip away the top layer of flux coating with a chipping hammer after you are done welding so you can inspect your work. The weld on the left in Figure 3-12 is a nice clean bead of weld with good penetration into the metal, but the weld the right is lumpy and rough, a sure sign that either you had the electrode tip too far from the metal, or the amperage setting was too low.

GETTING THE PROPER AMPERAGE

Experiment with the amperage setting on your welder as you practice drawing beads of weld on your scrap metal. When the amperage is too low, you will get a lot of sticking and false starts. Also, the flux will seem harder to chip from the weld. When the amperage is too high,

Figure 3-8 A smooth bead of weld on a test plate.

you will burn a hole into the metal after a few seconds similar to the one in Figure 3-9.

A good amperage setting allows you to weld continuously along the top of your metal without burning a hole or sticking to the work. Adjusting your amperage setting controls the amount of heat that is put into the work, and this is what makes a good weld. All the theory in the world won't teach you to instinctively control your heat; only a lot of practice will.

Continue to lay beads of weld from along the top of your work in a smooth straight line from one end to the other, and then start a new line along the last one, until you have created a raised surface on your practice piece. Have you noticed that the more you weld, the hotter the metal gets? Don't touch it to find out, just trust me here! Once you are able to weld thin bicycle tubing together, you will have to make frequent starts and stops in order not to burn a hole through the steel. This is because the tubing is so thin that it gets red hot in just a few seconds. Keep practicing on your heavier metal until you can lay down rows of beads, as shown in Figure 3-10.

Don't get discouraged if it takes a while to get good at striking an arc. The hardest part is learning to set your amperage properly and getting that arc going without sticking. This may take you all day.

WELDING PIECES TOGETHER

Once you are able to strike up an arc and lay some weld for an inch or two, it's time to move on and actually try to make two pieces stick together. Find two flat pieces of metal of equal thickness, and clamp

Figure 3-9 Excessive heat due to a high amperage level will result in a burn-through.

Figure 3-10 Laying beads of weld on the practice plate.

them together so there is a gap in between each piece about equal to the thickness of a welding rod (see Figure 3-11).

Make sure the ground clamp is connected either to one of the plates or to the clamp, or you will be striking all day with no arc starting. Also, don't weld too close to your clamp—you may hit it with the electrode or weld the work to it. Strike an arc anywhere on the work, and then bring the electrode into the joint between the plates, laying a slow, steady bead as you hold the electrode at a 45-degree angle from the work.

Keep your eye on the arc as you move along the joint, and watch to see if it is connecting to both plates. You may need to manipulate the welding rods back and forth slightly to get the arc to travel to both plates if it seems to favor only one side. Don't travel too fast along the joint or the final weld will be lacking sufficient filler metal from the rod. If you travel too slowly, a large crater will form between the plates, similar to the many holes you burned through the metal as you were practicing laying beads.

When you are done welding the entire length of the joint, remove the work and chip away the flux. You may also want to use a wire brush to clean the weld and surrounding area to get a better look at it (see Figure 3-12).

The weld on the left in Figure 3-12 is clean and solid, with adequate metal filling the joint. The weld on the right is lumpy and full of gaps where the arc stayed to only one side of the joint due to lack of heat or improper electrode angle. Keep trying this exercise until you can make

Figure 3-11   Two metal pieces ready to be welded together.

a smooth and solid weld between the two plates without gaps or holes. Remember, the key to making a good weld is to learn to control your heat by setting the amperage, and dragging the electrode at the appropriate speed. Are you running out of welding rods yet?

WELDING THIN-WALLED ROUND TUBING

If you have learned to strike an arc and weld two plates together, then it's time to take the final step and join some thin-walled bicycle tubing

Figure 3-12 A good weld produced using proper heat (left) and a poor one resulting from low heat (right).

together. This exercise will require some patience, so practice to get it right. Not only is round tubing the hardest joint to weld, it is also very thin, and this only adds to the complexity. Once you have mastered this next step, you will be a fairly good welder, but don't expect to get it right on your first attempt, or even on the first day.

Find a length of 1″ thin-walled round tubing without rust or paint covering it. A length of electrical conduit would be perfect, but any clean steel tube will do. Using a pipe cutter or grinder disk, cut the tube into several two- or three-inch lengths. Don't be too critical about measuring each cut, as you are just going to be using the metal for practice welding here.

Take two sections of tubing, and grind the end of one of them so they will fit together, as shown in Figure 3-13. Again, don't be too critical about making a perfect fit with the two pieces, as this will not always be possible, and a good welder can fill in the odd small gap. Remember to set up your ground clamp on either one of the tube sections before you try to strike up.

Figure 3-13   Tube joint ready to be welded.

Start with a lower heat setting to avoid a burn-through, especially if you have only welded the heavy practice plate. Bicycle tubing is very thin and it gets red hot in seconds, so don't expect to weld more than a half inch or so at a time.

Start from the top, and make a small bead approximately a quarter inch long just to join the two pieces together. Did you burn right through the tube? Even the most experienced welder sometimes blows holes in the work, and filling these holes is another handy skill to learn.

When you have a small bead of weld joining the two sections (see Figure 3-14), flip the work over and do the same thing on the top of the other side. Now the two pieces can be welded all around. Each time, only weld a small section no longer than half an inch, or you will burn through the tubing wall. Also, don't attempt to weld in any other position than from the top just yet, as this is something that takes a lot of skill to do properly. When you weld a small length then stop, it is a good idea to chip away the flux, so you can begin your new bead of weld slightly over the top of where you last left off, to avoid leaving a small hole in between the start and stop.

Figure 3-14   Start by welding the top of the joint only.

Keep turning the work around, welding small lengths at a time until you have made a complete joint, as shown in Figure 3-15. Don't worry about the appearance of the bead at this point. Try to avoid burning holes and leaving voids where you start and stop. Remember to hold you electrode so that it is at an equal angle between the two sections of tubing, not favoring one side or the other, or you will end up welding only one side of the joint. Remember, this is the hardest weld you will ever have to make, so take your time and practice until you get it right. If you practice these basic exercises, you will learn to make a strong weld.

GRINDING THE WELDS

Once you start welding your projects for real, it's a good idea to get into the habit of grinding finished welds. Not only does this make for a professional looking job, but it reveals areas that may need to be filled due to inadequate weld metal in the joint. Take a look at the two welds in Figure 3-16. Although both are equally strong, the weld on the left is rough looking due to many starts and stops and holes that needed to

Figure 3-15 Completed weld of two tube sections.

Figure 3-16 Rough weld (left) and smooth weld (right).

Figure 3-17 Grinding a weld improves overall appearance.

be filled. The one on the right was done very smoothly on the first attempt.

Now refer to Figure 3-17. Can you tell them apart? A weld does not have to be pretty to be strong and, once cleaned up with a grinder, even the roughest-looking weld looks good. When grinding a weld, avoid taking off too much material or you will weaken the joint. A ground weld should be flush or slightly higher than the material around it, and if you accidentally take too much off, add more weld to the joint.

If you would like to explore welding beyond the basics I have covered here, check with your local college about the courses available. A welding course will teach you many other aspects of welding such as "all position" welding; welding specialized metals such as chromoly, cast iron, and aluminum; tig welding; and mig welding to name a few. With patience and practice, you will learn to make a good weld, even if you know very little about the technology and terminology. So get at it, and burn some rod!

4

PLANNING AND DESIGNING

Planning Your Design

Before you start cutting up tubes and making the sparks fly in your workshop, it helps to have a design in mind for the project you plan to make. Your design can be as simple as a pencil sketch with one or two measurements or as complex as a CAD (computer-aided drafting) drawing with every single detail drawn with great accuracy and to scale. Having a plan helps eliminate those unforeseeable mistakes that only become evident in the final stages of the design, such as chain line problems or steering interference.

Of course, you could just start cutting up tubes and welding things together and produce a workable design, but this will happen only rarely and isn't worth the risk. Sometimes you may come up with a new idea or see a design you want to base your project on. Although your plan may seem to make sense in your head or look good from a photo, with careful planning, you might realize that there may be design flaws or hard-to-find custom parts needed, something the original designer may not have bothered to mention.

It's great to dream, and this is how a true designer gets ideas, but when it comes time to lay down the welding rod, be realistic when deciding which parts will be needed, and estimate the costs and total time needed to complete your project.

Let's say you want to build a recumbent trike, the type with two front wheels that steer and one drive wheel at the back. After looking over many designs and photos from the Internet, you decide that it doesn't look that difficult, so you begin to make the frame. The frame is a breeze because it only consists of a few pieces of round tubing and a few welds. Everything is turning out just as you planned.

Now comes the final stage of adding the steering and front wheels. Oops! You needed front wheels with 12 mm axels, and the cheapest set you can find cost more than $300. And what about those custom-machined axle holders for the Ackermann steering system? At $50 per hour for labor, it may cost another $300 to have those custom axel holders made at the local machine shop, even though the scrap metal is worth only a few pennies!

Keep in mind that your final design should not only include at least a basic drawing, but a cost estimate as well, especially if you are designing something that you plan to get some real use out of. Even a good set of high-pressure tires can run you over a hundred dollars for that low-racer or recumbent project, so think ahead.

The Power of Paper and Pencil

You don't have to be an artist to draw a bicycle or any type of machine; you only need to understand the object that you are drawing and be able to make a half-decent circle and some straight lines. Practice drawing a standard bicycle by looking at a photo or advertisement from a magazine and try to draw what you see. Don't worry about the artistic details such as shading, shadows, or logos; just keep to the important technical details that make a bicycle work.

Study the placement of the pedals compared to the rear wheel and ground, angle of the seat and head tube, fork length and spacing between the two wheels. When you draw a bicycle, it should look like it will work properly. Imagine the pedals rotating. Do they scrape on the ground or hit the frame? Can your machine steer properly? Many times, a real artist can draw or paint a really nice looking bicycle complete with shadows and reflecting chrome but, after a closer look, you notice that the geometry is incorrect or totally impossible. We want functionality here, not art. Save the artwork for your final design, when it comes time to paint.

When you can sketch out the basic bicycle on paper, you will now have true power. You can spend hours modifying the basic design into anything you like. This is how I come up with most of my project ideas, just by drawing many odd bicycles until something "clicks." See how little change is needed to convert a boring old mountain bike into a custom "raked" chopper (see Figure 4-1). It's much easier to plan your design and work out the costs or bugs on paper before you start investing time and money in something that may not even function.

Sometimes you might come up with a more complex design that has many details that cannot be seen from a simple side view, such as a multiwheeled vehicle or cycle-car. Have no fear; you can still get your ideas onto paper without the need to draw in 3D (something that requires great skill). Once you have drawn a simple side view, you can transfer most of the basic measurements to a top view by placing it directly over the side view so that both drawing line up. This type of drawing is done in the drafting trade to represent three views of an object on paper. This method can be very accurate and presents the final plan to the builder with much greater detail than any artistic 3D drawing ever will.

Refer to the recumbent trike plan in Figure 4-2. It's hard to get a good visual picture of how the steering will work from only the side view, but by transferring the basic measurements overhead to the top view, we can now see how it works. This method of drawing can also

Figure 4-1   A few simple changes turn a standard bicycle into a cool chopper.

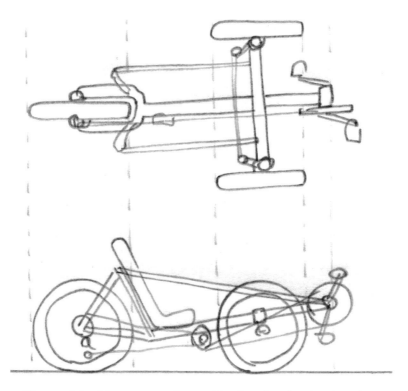

Figure 4-2   It's easy to project a side-view drawing to a top view.

work in reverse as well, if you started with an overhead view. All you have to do is draw light or dotted lines up from the important points such as wheels and frame joints and then fill in the blanks. Remember, accuracy is not the final goal of this type of sketch, only the overall picture.

Once you have drawn your design over and over from several angles until all the bugs have been worked out, you may then want to explore the world of computer-aided drafting or 3D rendering. Using a computer and a CAD program, you can input your measurements and create a drawing that is not only to scale, but totally accurate in every way. Figure 4-3 shows a rendered view of my Skycycle-4 project, a 12-foot tall impossible looking bike with the steering and chain hidden inside a 4.5-inch-diameter main tube made of electrical conduit. I really wanted to see how the unit would look when it was complete, so I put all of my measurements into the computer to generate an exact representation of the finished and painted bike.

Using a computer to create you design can produce amazingly accurate results, but it does take a lot of patience and skill with the soft-

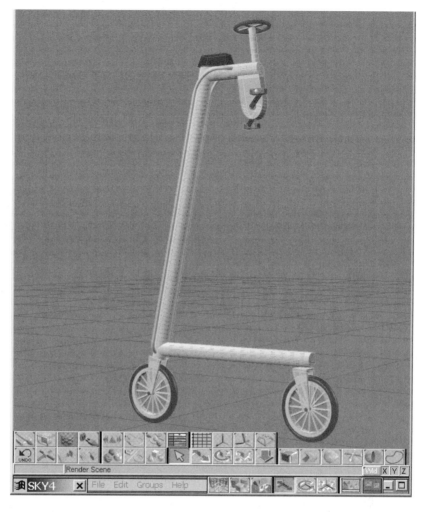

Figure 4-3 A computer-generated view of your creation can also help in the design phase.

ware program you are using. I would never start making a design on the computer before it was already drawn out on paper, as an eraser is a lot easier to use and rarely crashes. There are many computer programs that can be used to create a drawing or rendering of your project, from the very simple to the very complex (and expensive).

AutoCAD is a very popular drafting program in the professional world, but it does require a lot of skill to use and it's expensive.

The program I used to create Figure 4-3 is called trueSpace, from a company called Caligari Corp. (caligari.com on the Internet). There are many versions available, but I use an old demo version of release 2, available free as a download on their site. Although trueSpace does not have many of the functions of a full-fledged drafting program like AutoCAD, it does allow the beginner to start making simple 3D models like my Skycycle 4 within minutes.

So, get your pencil warmed up and start drawing bicycles until you can do it with your eyes closed. With the ability to quickly draw your designs on paper, you will find that your design stage really makes a difference in the outcome of the final product.

Designing a Strong Frame

Once you have the ability to put the ideas in your head onto paper, you will be able to work out a lot of your design challenges without having to break welds or redesign the working prototype. Of course, a frame that is structurally unsound on paper will become a structurally unsound frame in real life if you build it according to the plan. You can't expect a 10-foot single length of one-inch electrical conduit to support your full weight between two wheels, or a 12-inch kid's bicycle wheel to perform well at the high speeds demanded by a low-racer. There are certain mechanical limitations and rules that must be followed in order to make a successful project.

The first and most important principle to understand is that the triangle is the strongest possible structure that can be made. Look at any load-bearing structure and you will find that it is either a solid mass or made of triangles. Such examples are a typical slanted roof on a house, a bridge, a crane boom, and even the common bicycle.

The triangle is such a strong structure because for it to fail, one of the sides would have to bend. Look at the two structures in Figure 4-4. Imagine both shapes made by welding lengths of one-inch tubing together. As we push down on the tops of both structures, they will both fail in different ways. The square structure to the left will fail quite easily as the two joints at the left and right corners give out, collapsing the structure into a flat pile of tubes. The trianglular structure on the right however, will take an immense amount of force before any failure will occur. To cause a failure in the trianglular structure, the left and right sides must actually bend outwards, and this would take an amazing amount of force. Crushing this structure would be like trying to break a pencil by pushing down on one of the ends. Most likely, it would go through your hand before it would break.

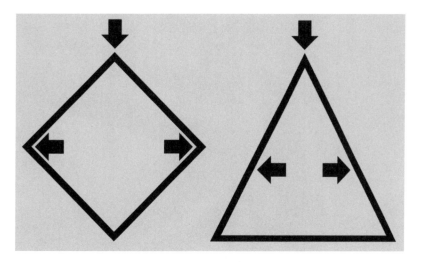

Figure 4-4 The triangle is the strongest structure that can be made.

If you look at a standard bicycle frame, you will notice that it is made of two triangles. One triangle is formed by the seat tube, top tube, and down tube, and the other by the seat tube, seat stays, and chain stays. This design has been around since the invention of the bicycle, even before they figured out that you could add a set of pedals!

Although not many of the bikes in this book could be considered normal or standard, they do adhere to many of the principals used in standard bicycle frame design. Even the Marauder low-racer with its single boom frame has a trianglular structure at the rear of the frame to support wheel. For a horizontal tube to carry any real weight, it would need to be fairly strong and heavy. This is why the triangle shape is used so much in designs where weight is a concern.

Sometimes, you may come across a frame that cannot be made from smaller triangles, or just needs some extra support. For these situations, the "gusset" comes to the rescue. This little piece of steel is welded into the corners of a structure to basically form a small triangle between the two pieces.

As shown in Figure 4-5, a small gusset is added to the front of a chopper frame where there would be a lot of stress from the weight of the rider. With a gusset in place, the frame can no longer fail at the joint; instead, the tubes would have to bend. Remember our failing square structure from Figure 4-4? If gussets were placed in all four corners, crushing the structure would be a lot harder and the sides would bend before the joints failed.

Figure 4-5   The gusset gives strength to the corners of a frame.

In early BMX-style bikes of the 1980s, it was common to see a gusset near the head tube. Although this was mainly done for looks at the time, it did increase the strength of the frame. Today's BMX is made from heavier and higher-quality tubing, and a gusset is rarely added.

If you are not sure if your frame will need a gusset, then add it anyway. It's better to overbuild during the prototype stage and redesign later. Having to remove an extra gusset from an overbuilt frame is better than eating the pavement at high speed when a structurally unsound frame fails!

Another factor that determines the strength of a frame or structure is the materials used. Although you will be mainly working with mild steel, there are times when you may want to build with lighter or stronger materials such as aluminum or chromoly. There are many advantages and disadvantages to aluminum and chromoly construction, and you should try to avoid using these materials until you have the proper skills needed to weld them.

Aluminum is much lighter than steel, but more difficult to weld. Aluminum is also a material that does not like to flex. If your design is not well thought out, and an aluminum joint is subjected to vibrations and stress, it will eventually fail. Anyone who rides an aluminum-framed bicycle will tell you that it is much stiffer and unforgiving over a rough road than a steel-frame bicycle.

Chromoly is a special type of steel that has more strength than ordinary mild steel, so tubing made from it can be thinner and lighter. To weld chromoly properly, a pre- and postheating process is necessary in order to release the stresses created by the welding process. In a production environment, a large oven is used to heat the newly welded frames until they are red hot, something that is hard to do in a small garage. If chromoly is not welded using proper techniques, the resulting joints will be weaker than if regular mild steel were used.

Another material that is being used in bicycle frame construction is called carbon fiber. This is a very light and strong material that is formed in a similar manner to fiberglass. Other than the fact that it is amazingly light and strong, I know very little about using carbon fiber, and will leave it to the experts. There are people who have built their own carbon fiber frames from scratch, but this is not something I would recommend to a beginner.

If you take your time in the design phase, adding strength to the frame where it is needed, creating a strong frame is not a difficult task. Remember, if you don't think it will be strong enough, overbuild in the beginning, then try to make things lighter after a successful test.

Get out your sketchpad and start cooking up some imaginative designs so you can begin the process of turning them into real-life creations. You have almost everything you need to start building—a source for parts, basic tools and skills, and the ability to plan ahead.

In the next chapter, I will show you what each part of a bicycle is called and how to remove it for use in your own projects. Are you ready to get your hands dirty?

5

TAKING IT APART

Over the last 150 years, the bicycle has evolved from the clunky wooden "velocepede" to the fast, lightweight, high-tech machine of today, but the general geometry and design have remained the same with very little change. The frame geometry is still made of two triangles with the rider placed above the center of the frame. This design is very strong, gives the rider a safe vantage point in traffic, and offers an easy way to bail from the bike in an emergency. Although the bike projects presented in this book are completely radical and off-the-wall, many of them, including the choppers and tall bikes, actually follow the same basic trianglular frame. Taking a complete bike apart right down to the frame is not very difficult, and requires only basic tools.

Every Piece Has a Name

In this chapter, we will be taking apart a common mountain bike (Figure 5-1) down to the frame. Although there are hundreds of different sizes and standards for bicycles and their components, the basic disassembly procedure is fairly standard. However, before you start breaking down a bike, it is a good idea to get to know the names of the various bits and pieces, so when you have to ask for something at a bike shop, you will know exactly what to ask for, and when I tell you to chop out the head tube, you will know where to aim your hacksaw.

BASIC BIKE PARTS

In reference to Figure 5-1, each part of the bike has a name:

A   **Tires**—Some common sizes are 20, 24, 26, and 28 inches.
B   **Rims** come in a range of diameters and widths.
C   **Valve Stem**—A one-way valve to fill the tube with air.
D   **Dropouts**—Hold the hubs in place by tightening a nut on the axle.
E   **Freewheel/Rear Sprocket**—A single or cluster of gears to drive the rear wheel.
F   **Rear Derailleur**—Allows the changing of the rear gears by moving the chain.
G   **Seat Stay**—Part of the rear frame from the seat to the rear wheel.
H   **Chain**—These come in various widths depending on the usage.
I   **Chain Stay**—Part of the frame from the crankset to the rear wheel.
J   **Brakes**—Stop the wheels due to friction caused by clamping a shoe against the rim.
K   **Seat**—Connected to the adjustable seat post, which is held in place by a clamp.
L   **Seat Post**—Can be adjusted up and down and locked in place by a clamp.
M   **Seat Tube**—Part of the frame between the seat and the crankset.
N   **Front Derailleur**—Allows the changing of the front gears by moving the chain.

Figure 5-1 It is important to understand each part of a bicycle and how it works.

O **Crankset**—The combination of front sprockets, crank arms, and axle.

P **Pedals**—Transfer power from your legs into the crankset.

Q **Top Tube**—Part of the frame between the forks and the seat.

R **Down Tube**—Part of the frame between the forks and the crankset.

S **Brake Levers**—Transfer force to the brakes through a cable system.

T **Handlebar**—Steering system of the bicycle. Brakes and shifters are mounted on it.

U **Gear Shifters**—Transfers motion to the derailleur, allowing the changing of gears.

V **Gooseneck**—Connects the handlebar to the forks. Allows for height adjustment.

W **Head Tube**—Joins the top tube and down tube, and contains the forks and bearings.

X **Forks**—Connects to the head tube and holds the front wheel to allow steering.

LEARNING BY DOING

If you are going to be fixing or building bikes as a hobby, it is a good idea to understand how everything all works and fits together. Find a complete working bike and study how each part works with the others. Turn the bike upside down, turn the cranks, and watch how the derailleur moves the chain up or down the gear cluster as you shift. Turn any adjusting screws and see how the adjustments affect the way things work. Experiment with the bike and learn how it works by operating it. The best way to learn is to jump right in and give it a try.

There is really no reason to take the smaller pieces like brakes and derailleurs apart because very little inside can be fixed or adjusted anyway.

COMMON PROBLEMS AND HELPFUL TIPS

If a brake arm is bent or a derailleur wheel is cracked, just toss it out and find another one. You won't be working with really expensive components here.

A common problem with older bikes is bad cables. Weather causes many cables to become seized or rusted, and these need to be cut off and tossed out. When you get a bike for parts, try the shifters and brakes to see if they move. If they seem to be seized up solidly, then cut all the cables with a good pair of wire or bolt cutters and toss them away.

In Figure 5-2, you can see the four main cabled components: front brake, rear brake, front derailleur, and rear derailleur. It is a good idea to get familiar with the proper workings of these devices, as they are essential for most bike projects. It's not a good idea to test drive your first low-racer without brakes. Trust me, I know from experience!

You will find that the smaller details, such as cables, shifters, and brake levers (Figure 5-3), and derailleur alignment and brake pads take a lot of time and effort to get right. Try to avoid cutting corners, especially with brakes. Out of a dozen old scrap bikes, there may only be one usable brake, so know what to look for.

If you find yourself short on good cables, any bike store can cut you off a length for a few dollars, or you can buy precut lengths at most hardware stores that also sell bikes or bike parts.

BEARINGS AND THREADS

Another good bike building skill is to know your bearings and threads. On a typical bike, there will be ball bearings in almost any part that

Figure 5-2 Make sure that all cables, shifters, brake levers, derailleur alignment, and brake pads are in good working condition.

moves (see Figure 5-4). The forks require two, the crank requires two, each pedal requires two, the rear wheel has four sets, and the front wheel has two. Each bearing set has a *cup* or *race* that it rides in, and many of these are threaded in place to another part or the frame. These threaded parts will contain both left-handed and right-handed threads.

It is really tough to remove the right-side bottom bracket cup if you are trying to unscrew it in the wrong direction. As you take a bike apart, really take a good look at which way all the parts come out, and take note of thread size and direction. You will save yourself a lot of time and broken tools by knowing how to remove a part with the least amount of force.

Bearings only fit into the cup in one direction as well, so watch carefully as they are lifted out. Generally, a bearing set has a *flat side,* and

Figure 5-3 Brake and shifter levers.

Figure 5-4 A complete one-piece crankset.

a *balls side*. The *balls side* goes into the cup. If you do get it backward, it's fairly easy to tell, as you will get a lot of resistance from friction or a scratchy sound as metal grinds on metal. When you have removed a part with bearings, try putting them in backward and see how they work. Chances are, they won't work properly.

The forkset bearings (see Figure 5-5) come in many sizes depending on the bike, so you may want to put a twist tie around matching sets. The difference in size is so small that it is hard to find the right size you need if all your bearings are thrown in a large pail. If you put too small a bearing into a cup, it will not work and will be damaged.

Another feature to notice about the fork bearings is that the lower cup may be a little deeper than the top. This is because most of the

Figure 5-5 Forkset and bearings.

weight is pushing down on this set during normal riding, so it needs to be a little heavier. The actual bearings are the same size.

Also, BMX head tubes are normally heavier than regular bikes, so the bearing cups from the two are not usually interchangeable. A head tube cup should fit snugly in place. If there is excessive slope, the bike may start to wobble and shake when moving at high speed. A good fit requires a hammer to bang it in place, but if the cup is too big, it will not fit.

Once you have taken apart a few bikes and have a collection of parts to work with, you will notice that many parts can be interchanged between different bikes. You can swap chain rings, forks, and handlebars, and create some interesting hybrids without even cutting or welding the frame. Do you want a speed bike with massive, knobby tires? A small BMX bike with 36 gears capable of high speeds? Sure, why not? Anything is possible.

Don't be afraid to grab your toolbox and dig right in. This is a great hobby and is easy to learn. Although many of the bike projects in this book can ride remarkably fast or are designed mainly to look cool, no special skills or knowledge beyond the basics are needed to make them from a pile of scrap bikes. If you want to take bike building to the extreme and learn to spoke your own wheels or drill and thread your own hubs, that's great, but none of that type of skill will be necessary for any of the projects presented here.

Stripping it Down to the Frame

OK, now that you can talk the talk, let's pull it all apart and make a big mess on the workshop floor. The best place to begin is by taking off the wheels so you can work with a smaller package up on your workbench.

TAKING OFF THE WHEELS

Grab an adjustable wrench and begin by loosening the rear nuts in the counter-clockwise direction five or six turns, as shown in Figure 5-6.

If there is a lot of rust on the axle, you may need to use two wrenches, one on each side so the entire axle doesn't spin around, as this will destroy the threads. If this is the case, then make sure the wrench is tight or use a socket set to avoid stripping the nut. If the nut is already stripped or worn round, you will have to use a vice-grip, or pipe wrench to get it loose. When using vice-grips, try to avoid crushing the threads or you will be only making the problem worse.

Figure 5-6 Removing the rear wheel nuts.

Once you have the two rear axle nuts loose, pull back on the derailleur so the wheel can just drop out. Depending on the tire size, you may also need to either let out some of the air or loosen the rear brake cable so the tire can squeeze between the brake shoes (see Figure 5-7).

Mountain bikes with fat, knobby tires will have this problem, but letting out some of the air is usually enough to allow the wheel to come out with a little force. Removing the front wheel is even easier—just loosen the two nuts, and it will drop right out. Some front wheels have special washers that hook into the forks, so you may need to loosen the nuts a fair amount in order to get them to come loose from the slots. These special washers keep your front wheel from falling out in case the nuts become loose while riding.

Now that you have the wheels off, you can work on the bike atop your workbench, since it takes up less space now and won't roll right off. If you are planning to strip the bike right down to the bare frame, this is a good time to remove the chain. This will allow the derailleurs to be removed without taking them apart.

Figure 5-7 Removing the rear wheel.

REMOVING THE CHAIN

The best way to remove the chain is with a chain link tool. This can usually be purchased for under $20 from any bike store and will last a lifetime. Place the tool on a link and turn the crank as shown in Figure 5-8. The link will pop out with little fuss. Doing this procedure with a hammer and punch is not only frustrating and slow, but can sometimes damage the link or your finger when you miss. If you do decide to do it the hard way, put a small nut under one of the link pins, then bang the top with a hammer until it is flush with the link. Now take a small punch or nail and bang the pin until it is almost all the way out so you can separate the links.

Once the link is open, pull the chain through both derailleurs and set it aside. If the chain is very rusty, you can try soaking it in a can of Varsol™ or oil, but if it seems too far gone, don't waste your time; throw it away. After soaking for a few days, the links should move easily.

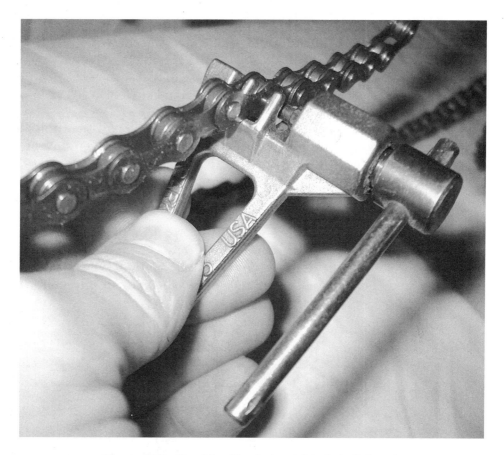

Figure 5-8   Breaking the chain with a chain link tool.

Now it's time to get rid of the small stuff—brakes, derailleurs, and cables. At this point, you should decide if you want to keep the cables or just cut them right off.

Many bikes found in scrap piles or yard sales may be rusty and seized, so cutting cables with side cutters may be a lot faster than removing them the "nice" way. Try the brakes and shifters. If they do not seem to move at all, just chop them off and send them to the garbage heap. If they seem like they may still be useful, then let's start removing all the cable ends from the brakes and derailleurs.

REMOVING CABLES

Brake cables are usually held on by a hollow bolt that crushes the cable against the brake arm. To remove the cable, loosen the nut until you can pull the cable right through the hole (see Figure 5-9). If there is a small cap crimped to the very end of the cable, pull it off with pliers. This is used to stop the cable from fraying, and cannot be put back on later. Also, be careful with frayed cables because they can cut you easily if they are pulled with bare hands. Wearing a pair of work gloves is a good idea here.

Once you have both brake cables out, pull out all the cables so they are only connected right at the brake levers. Cables may be routed through holes or knobs in the frame, and it may be necessary to pull the sheaths right off the cables in order to get them all the way out of the frame. Once you have the brake cables done, do the same for the front and rear derailleurs, as shown in Figure 5-9. Start by loosening the nuts that hold the cable ends and pull them all out right back to the handlebar or frame-mounted shifters.

In Figure 5-10, you can see the small cap that is crimped on the cable end. This should easily pull off with a pair of pliers and is of no

Figure 5-9 Removing the rear brake cable.

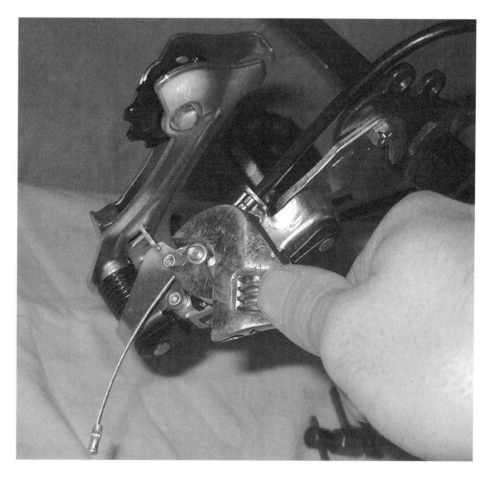

Figure 5-10   Removing the derailleur cable.

use once removed. There should be no cable attached to the bike except for those at the shifters and brake levers mounted on the handlebars.

REMOVING THE HANDLEBARS

Now is a good time to remove the handlebars, so that all of the stray cables are out of the way. The fastest way to get them off is by removing the gooseneck rather than all the levers and grips. It all comes off by loosening one bolt—the one in the middle of the neck. This may be a standard bolt or require a hex key wrench (see Figure 5-11). Either way, find the size you need and turn the bolt counterclockwise about 10 or 15 turns.

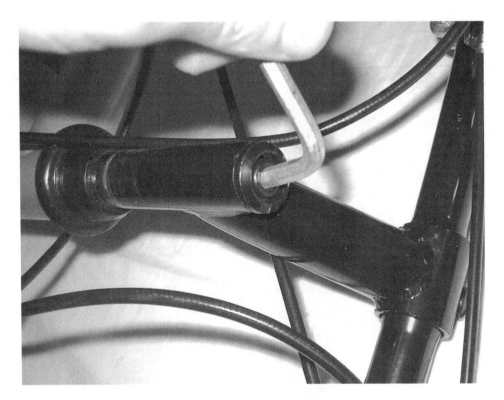

Figure 5-11   Removing the gooseneck and handlebars with a hex key.

If you take this bolt out all the way and try to force out the neck, it won't budge, even with a large hammer (I tried this once). The reason for this is the wedge-shaped nut at the end of this bolt. As the bolt tightens up, the bolt gets wedged against the side of the fork tube, holding the neck firmly in place.

Once the bolt is out about half an inch, tap it down with a hammer so it is again flush with the top where it started. Now you should be able to turn the neck back and forth until it comes out of the fork tube. If the bike is rusty, this may take some work, but remember not to take this bolt all the way out.

As shown in Figure 5-12, the wedge-shaped nut only needs to be pushed down half an inch or so in order to release the force against the fork tube. If the bike is in really bad shape, you may need to secure the forks and put a lot of effort into cranking the handlebars back and forth until they come all the way up. Always use the hammer as a very last resort, but do whatever it takes to win!

Figure 5-12 The gooseneck and its locking nut.

REMOVING THE BRAKES

To remove the brakes, undo the small lock nut on the back side of both brakes, as shown in Figure 5-13. The front and rear brakes are the same, so the operation will be identical for each.

If there is a lot of rust, you will need to keep the nut on the other side from turning. Once the brakes are off the frame, make sure they still work by pressing them together. If they are bent or badly rusted, it is recommended that you find a better pair, rather than finding out at the last minute that they have failed.

The rear derailleur is usually held to the frame by the wheel nut and one small bolt for alignment with the rear dropout. It is this alignment nut that will be holding it on, since the wheel is no longer attached to the bike (see Figure 5-14). Just turn the bolt a few turns and the derailleur will fall right out.

REMOVING DERAILLEURS

Some better-quality bikes have the derailleur mounted directly to the frame using a special dropout. If this is the case, leave it with that frame; it will not work on any other. Once the derailleur has been re-

Figure 5-13   Removing the brakes.

moved, inspect the unit for spring tension and spin the little plastic wheels to make sure none of the teeth are missing. Unwrap anything that may be tangled around the axles, impeding their movement.

If you have not yet removed your chain, take one of the small wheels off the derailleur by removing the bolt that holds it in place, and then put it back together after the chain has passed through the opening.

The front derailleur is the last bit of small hardware that needs to be removed from your soon-to-be bare frame. Most of these are clamped to the frame with a small bolt. This bolt needs to be removed all the way in order to allow the clamp to let go of the frame tube (see Figure 5-15).

Once open, the unit should slip right off after spreading apart the two halves of the clamp. If your chain is still connected, just remove

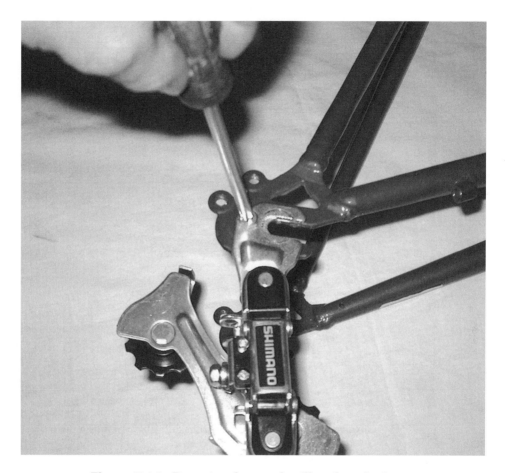

Figure 5-14 Removing the rear derailleur from the frame.

the small bolt at the back of the derailleur and slide the chain through the opening. If your derailleurs are heavily rusted, drop them in a bucket of Varsol™ or some other metal-cleaning agent. Check for bent arms and seized springs as well.

REMOVING THE SEAT

The seat and seat post are the next to go. This job is either going to be really easy or really hard, depending on how much rust is in between the post and frame.

Leave the seat on for now. This is what you are going to hang on to as you twist the post out. Remove the bolt from the clamp on the frame and slightly spread it apart with a flat-head screwdriver, as

Figure 5-15 Removing the front derailleur from the frame.

shown in Figure 5-16. At this stage, you should be able to hold down the frame and twist the seat and post out of the frame. This may take a lot of effort and some time if it is a long seat post or full of rust. If it seems to be very stuck, with no movement at all, it may be time for the vice grips or pipe wrench and, even then, it may be a chore to remove it.

If all of this fails, remove the seat and crush the top of the post in a large bench-mounted vice, and try to crank it out by holding onto the frame. If this still fails to loosen the post, you are on your own. Get a large hammer!

REMOVING THE FORKS

Removal of the forks is straightforward, but you will need a large adjustable wrench or pipe wrench. Always use an adjustable wrench first, as it will not leave teeth marks in the soft metal parts. All

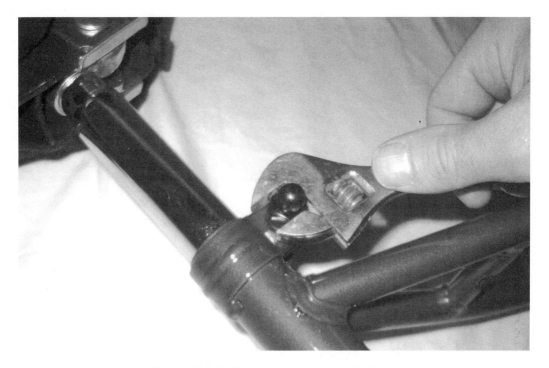

Figure 5-16 Loosening the seat-post clamp.

threads on forks are right-handed, so turn the top nut counterclockwise until it is completely off (see Figure 5-17).

Under the first nut, there will usually be a nonthreaded ring with a notch in it. This ring will lift right off. The next ring is the top of the bearing race, and this will also unscrew, allowing the forks to be removed from the head tube (see Figure 5-18). Collect the rings and bearings, and store them in a safe place. If the bearings fall all over the floor, this means that the ring that holds them is cracked or rusted away, so don't bother trying to collect them all; this bearing set is finished.

When the forks are removed, inspect them for cracks and bends. Check the threads and dropouts to make sure there is no major damage.

If you have trouble removing the second ring, it may be because the small notch that stops it from turning has been forced out of the hole and into the fork threads. This is fairly common and can be easily fixed by grasping the ring with vice grips and carefully turning it until it goes back into place. Once in place, the ring should easily lift out without force.

A reflector mount that has been forced out of place could also be responsible for damaging the fork threads, but as long as the majority of

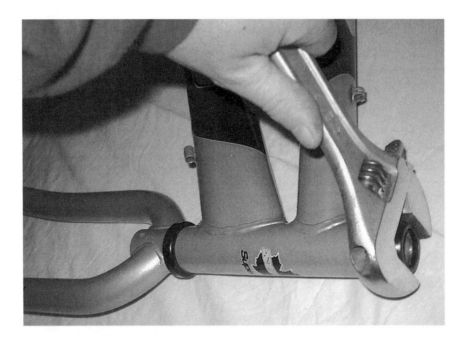

Figure 5-17   Remove the top nut to begin disassembling the forks.

Figure 5-18   Removing the bearings from the forks.

the threads are still good, the forkset will be fine. If all the threads are gone, the forks should only be used as parts for other projects. You don't want to lose a front wheel on a ride.

REMOVING THE CRANKSET

Now you are down to the last and most complex part—the crankset. First, you need to take both pedals off. This may be the toughest part to do if rust has eaten into your bike. You will need an open-ended fixed wrench, usually 5/8-inch, most likely a good hammer, and possibly even a blow torch.

The most important thing to remember is that the left pedal (opposite the chain ring) has a left-hand thread. If you want to loosen it, you turn it clockwise, as if you were tightening it. A lot of pedals are stamped "R" and "L" on the inside (see Figure 5-19), just in case you forget which one is the oddball.

Figure 5-19 Pedals have a left and right side, as shown in this picture.

Now that you know which way to turn the wrench, give it a try. If your bike is clean and fairly free of rust, it may just come right out, but this is not usually the case. In fact, I have had to cut several crank arms right off with a grinder, even after several attempts at heating it red hot and hammering on the wrench.

You may want to keep a good stock of 5/8-inch wrenches around if you plan on taking a lot of old rusty bikes apart. You will almost always have to bang on the wrench with a hammer, and a lot of the time, you will snap one end of the wrench. No kidding! Hopefully, your pedals came off with ease, but if not, it's time for the heavy artillery.

Get your biggest hammer and bang the wrench in the proper direction (see Figure 5-20). Remember, right side = right threads, left side = reverse threads. If the pedal will not budge, you can apply some heat using a propane blowtorch or welding torch. Heat up the area around the pedal, not the pedal itself. Heat makes metal expand, so this may loosen the grip on the pedal threads. If your pedal is plastic, it will probably melt during this operation, but this cannot be avoided, as you need a lot of heat for this to work. Ten solid minutes with the torch is about right.

Figure 5-20   Using a wrench to remove a pedal.

If all of your attempts fail, you may end up cutting of the crank arm with a grinder, but that is still better than nothing. The frame and chain ring will still be intact.

Crankset Types

Now that you've removed the pedals, it's time to pull the crankset from the bottom bracket (the ring on the frame that holds the bearings and cups). There are two types of cranksets—one-piece and three-piece (see Figure 5-21). One-piece cranksets are common on BMX type bikes and cheaper hardware store bikes. They have a continuous "S"-shaped crank arm made of solid steel. This type of crank is good for budget projects or projects in which weight is not a consideration.

As you can see in Figure 5-21, the bottom bracket for the one-piece crank (right) is quite a bit larger than for the three-piece (left). Also, the three-piece crank bracket is threaded on both sides, whereas the one-piece crank bracket has no threads at all.

Three-piece cranksets are found on racing bikes and bikes of medium to high quality. There is an axle with squared sides connecting the

Figure 5-21 Bottom brackets. A three-piece crank (left) and one-piece crank (right).

cranks on each side of the bike. The square sides of the axle are slightly tapered so, as the end nut is tightened, the crank arms get wedged
on very tightly. The axle is made of hardened steel and the crank arms
are made of steel or aluminum. This type of crank is good for projects
requiring light weight, precision, and high gear ratios. The first type of
crank we will remove is the typical three-chain-ring, one-piece setup.
The nice thing about removing this crankset is that only two bolts hold
it together.

Crank Removal Tips

Before you try to remove the crankset, make sure the left-side pedal
has been taken out, or the rings will not slide off. If you can't get the
pedal out, it's time to get out the grinder and cut the arm off or you will
never be able to remove the crankset. Find a large wrench and turn
the large nut clockwise (another backward thread). This is shown in
Figure 5-22. It should come off with little effort unless you tried to
turn it the wrong way.

Figure 5-22 Removing the locking nut from a one-piece crankset (with reversed
thread).

Underneath is a ring or washer, and it should lift right out. If it is stuck, it may need to be turned until the small notch is aligned with the small keyway in the threads. Once the washer is off, there is only the bearing race left, and it may come off just by pushing it with your fingers.

Don't forget, the threads are backward, so turn it like you're trying to make it tighter. If this ring seems to be a little stuck, find a punch or old screwdriver and place it in one of the slots. Then tap it with a hammer as shown in Figure 5-23. Once you have it moving a little, it should easily come all the way off.

If the threads are very rusted or damaged, you may need to bang the ring all the way off, but this is pretty rare. Be careful not to smash the crank arm threads, or it will be hard to get off. Once the last ring is free, remove it along with the left bearing. If the bearing is wedged in a clump of dry grease, pluck it out with a small screwdriver, but try not to bend the thin ring that holds the balls in place (see Figure 5-24).

Figure 5-23 Tap a screwdriver into the slot located on the ring in a clockwise direction.

Figure 5-24 Removing the bearing and ring from inside the cup.

If the bearing ring is rusted and cracked, and the balls fall out, discard them. It is important to take this bearing out or the crank arm will not be able to slide out through the bottom bracket.

Once the left bearing is out, slide the entire crankset out by aiming the chain ring into the middle of the frame while guiding the crank arm through the two bearing cups. As shown in Figure 5-25, as long as nothing is bent or damaged, the crankset should come right out without having to force it in any way.

Place all the bearings and rings in a safe place or bucket of solvent if they need to be cleaned.

The Three-Piece Crank

Removing the three-piece crank is a bit more involved, as there are a few more parts to deal with, and things can get really hard to separate, depending on the amount of rust or age of the bike. This time you will definitely need your hammer.

Start by removing the plastic covers in the center of each crank arm, if there are any, and you will find a 14-mm nut at each end (see Figure 5-26).

Figure 5-25 Sliding the one-piece crank out of the bottom bracket.

Figure 5-26 Crank arms are fastened to a shaft with nuts.

Clean out any dirt from the recessed area, and spray a little rust remover in the hole if there is a lot of corrosion inside. To remove these nuts, find the correct size ratchet and turn them in the counterclockwise direction, as shown in Figure 5-27. Do not try using a wrench here. The holes are recessed too far and you will only strip the end of the nut. If the nut is really stuck, hold on to one of the crank arm or tie it to the nearest part of the frame so it will not move.

Taking out these nuts is not normally hard to do unless the threads have been damaged or flattened at the end. In this case, you may be fighting a losing battle. Once you get the two nuts off, the real fun will begin as you attempt to remove the two crank arms. This is probably the most frustrating part of taking apart an old, rusty bike, as these two parts have had years to jam into place. Before you start, take a look at how the crank arm is connected to the axle (Figure 5-28). The square cutout is slightly tapered so, as the nut turns, it pushes the crank arm very tightly against the axle.

Figure 5-27 Remove crank nuts by turning them in a counterclockwise direction with a ratchet.

Figure 5-28   A square tapered shaft holds the crank arm.

Removing the Crank Arm

Even on a brand new bike, a fair amount of force is needed to free the crank arm, so be prepared. If the arm is aluminum, it may come off fairly easy due to the softness of the metal, but you are still going to need a hammer for this operation. Start with the sprocket crank arm first. This will allow you free access to the other arm once it is out of the way. Place the bike up on the workbench so the chain ring is hanging off the end. It should fall to the ground when you finally set it free. With the bike hanging over the edge of the bench, place a solid rod or wedge-shaped piece of steel against the crank arm through one of the openings in the chain ring, as shown in Figure 29.

Place the tool against the crank arm, not directly against the chain ring itself, as the ring will bend from the force. An old socket extension also makes a good tool to bang the crank arms off, but make sure it's an old one because you will be doing a lot of hammering.

Figure 5-29   Some force will be needed to remove the crank arms from the shaft.

If the crank arm is refusing to let go, you may need to get out the torch and put some heat into it first. Apply heat for five to ten minutes all around the middle edges of the crank arm, but not right into the recess where the axle bolt is. The idea is to expand the crank arm away from the axle, and heat makes metal expand. Once you get this arm off, the other one should be a little bit easier, as there is nothing in the way (see Figure 5-30). Use the same technique as before, applying heat if needed.

Figure 5-30 With the chain ring out of the way, the left side is easy to remove.

Set your hammer as close to the axle as possible, but be careful not to smash the end of the axle and flatten the bolt. If this all seems to be too much work, you can buy a crank puller, but it will cost you a lot more than a new set of cranks, and may not be able to remove the old rusty ones. Once you manage to get both the arms off, you can now remove the axle and bearings from the frame for cleaning or replacement. For this job, you will need either a pipe wrench or a large adjustable wrench. Start with the left side of the frame. This will be the side with the large notched ring over the smaller threaded cup, as shown in Figure 5-31.

Grasp one of the notches with the pipe wrench and turn it counterclockwise until it comes loose, then it should unscrew the rest of the

Figure 5-31 Removing the lock nut with a pipe wrench in the counterclockwise direction.

way with little effort. This is the retaining ring that stops the bearing cup from coming loose as you pedal (see Figure 5-32). Once the retaining ring is gone, you can now remove the inner ring in the same manner. Grip the flat sides of the ring with your wrench and unscrew in the counterclockwise direction until it is all the way out.

If the ring is badly corroded, it may be very difficult to get it out. Again, you could try using the torch by heating up the underside of the bottom bracket. It is best to use a wrench that fits snugly on the ring, as it is very easy to strip the edges from excessive force. If all else fails, use a punch and try to hammer the ring into turning enough to get it going with the wrench. If you have to resort to this, make sure you use the torch first. As shown in Figure 5-33, once the ring is out, you can pull the axle and both bearings out for cleaning.

Notice the direction the bearing is facing—balls up, ring down. If you try to put it back together in reverse, it will not work. If bearings fall all over the bench, this means the rings have corroded and the bearings are no good. Check the rings for cracks or excessive wear, and

Figure 5-32 Removing the bearing cup using an adjustable wrench in a counter-clockwise direction.

Figure 5-33 Bearings, cup, and axle can now be removed.

toss them if they show any signs of damage. Also check the threads in the bottom bracket.

All that is left is the drive-side ring, and for this you will need your large wrench or pipe wrench again. Take note that this side is reverse threaded; you will be turning it clockwise to loosen it (see Figure 5-34). Sometimes, this ring can be really stuck due to rust and dirt, so you may need your entire army of tools—wrench, hammer, blowtorch—to get it out.

You will notice that there is not much of the ring above the bracket. Make sure the wrench is set perfectly or you will slip or round off the edges. A little heat directed at the underside of the frame can sometimes make a lot of difference in this battle. Apply the torch for five or 10 minutes, then try it again.

Inside the Bottom Bracket

Figure 5-35 shows the parts that make up the bottom bracket—the axle, two bearing rings, a lock nut, and two bearings. Close inspection

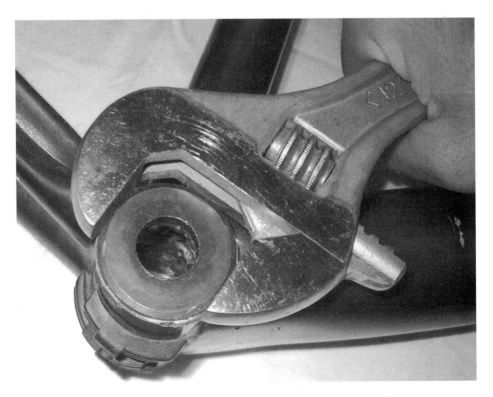

Figure 5-34 Removal of the right-side bearing cup in the clockwise direction.

Figure 5-35 Parts that make up a three-piece crankset.

of the axle will reveal that one side is slightly longer than the other. The long side is the side to which the chain ring attaches, and the extra length is to help the chain ring clear the frame.

The cup without the lock nut is also on the chain-ring side and has a reverse thread, and the one with the lock nut is on the left side of the frame with right-hand threads. Bearings always face into the cups so that the balls are in first and the flat side is against the axle (see Figure 5-36). If you are not sure which way to put bearings back together, just try it. One way will have low friction even if you press as hard as you can. The other way will result in metal against metal, and it will feel a little stiff or harder to turn as you push on it. Now, the only parts left on the frame should be the bearing cups.

Find a rod or piece of metal long enough to reach into the head tube and get your hammer ready. Place your rod or screwdriver down the tube so it catches on the edge of one of the cups (see Figure 5-37).

Figure 5-36   Bearings only fit one way into the cup (balls down).

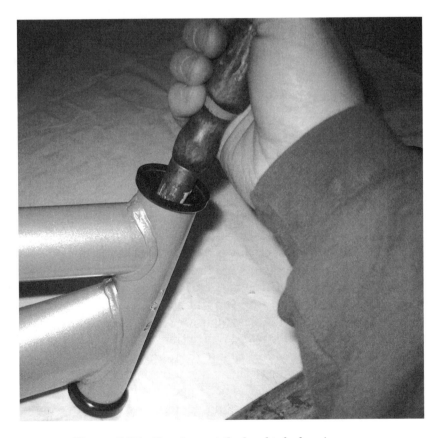

Figure 5-37   Banging out the head tube bearing cups.

Because they are just pressed to fit, a few whacks with the hammer should get them out with little effort. The same process is used for the bottom bracket cups as for the head tube cups. If you have an old seat post lying around, this works great for hammering out the bearing cups. Most bottom bracket cups are the same, but fork bearing cups come in many different sizes. BMX and heavy mountain bike frames have larger head tubes than light racing frames and kids bikes.

The cup should require a hammer to put it into the frame. If it just slides in with little effort, then it is probably too small. Of course, you shouldn't have to smash it in with a sledgehammer, either!

GETTING ORGANIZED

Congratulations, you have taken a bike apart right down to the frame. If you plan on collecting a lot of bikes for your "scrap pile" then this is the best way to organize the parts. Separate all like parts such as pedals or cranks into large plastic tubs, and put bearings, chains, and small parts in a tub with Varsol™ or other metal cleaner until they are free of rust and grime. Bikes take up a lot less space when broken down, and you can even hang all the frames on a wire, like clothes on a clothesline, out of the way.

Chopping up the Frame

Not every bike you acquire will be in good shape. In fact, unless you plan on paying top dollar at yard sales and auctions, most of the bikes you get for free will be bent, rusted, worn, dented, and generally beat up. But the best part is that you can get some decent frames and parts for free!

The good news is that, most of the time, the good parts will still be intact, even after the bulldozer at the local landfill site mows the bike into a 20-foot-high twisted heap of scrap metal.

Many of the projects in this book only use part of the frame, and some use none of the frame tubing at all, only pieces like the bottom bracket, head tube, rear dropouts, and other parts that you generally can't make easily. When you can buy a 10-foot length of 1-inch, thin-walled electrical conduit for $5, why mess around with rusty, dented frame tubing? Sometimes I'll find an old bike at the scrap yard and chop it up right on the spot. I rarely use the frame tubing in my work.

USING A HACKSAW

Now that you see what I am getting at here, let's refer to Figure 5-38. Each number shows you the best order in which to chop a frame to bits. Numbers 1, 2, 3, and 4 actually require two cuts each. Since there are two legs to each part, you will cut them both.

It is a good idea to hold the frame securely in a vice or clamp it to the workbench so it's secure while you cut. It is very easy to snap a hacksaw blade if the work you are trying to cut turns and snags the blade (see Figure 5-39).

If you are using a grinder with a cut-off wheel and your frame is not securely held in place, this can send the frame or your grinder flying across the room. Also, when gripping the frame tubes in the vice, don't overdo it and crush the thin walls, unless you plan on discarding them anyway. Bicycle tubing is fairly thin and will flatten easily.

You can use a hacksaw or a grinder to make the cuts, but try to keep each cut as close to the joint as possible. This will ensure that the maximum length of frame is available, and it will be much less work to grind the stubs from the bottom bracket and head tube, especially if you plan to use a hand file to clean up the leftover metal (see Figure 5-40).

Figure 5-38   Cutting order for chopping up the frame.

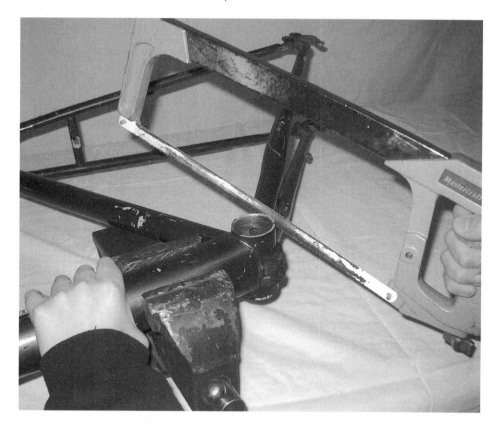

Figure 5-39 Cutting is easier when the frame is held securely in place.

I've suggested a numerical order for the cuts (refer back to Figure 5-38) so you will be able to get your blade as close as possible to the tube joint without sawing in an awkward position. If you attempted to cut the seat tube first, the saw would hit against the chain stays or down tube and prevent a flush cut.

Hacksawing is an art in itself and, once mastered, you will be able to cut off a tube right down to the base in a straight, even line without any problem at all.

HACKSAWING AND GRINDING TIPS

One thing to avoid when using a hacksaw is "speed cutting." One stroke per second is a good pace to use. This method will allow you to guide the cut into a straight line or along a marked path. If you saw away like crazy, the cut will be crazy or you will snap the blade very quickly.

Figure 5-40   Cutting close to the joint prevents excessive filing or grinding later on.

When using a cutoff wheel for your grinder, be careful when you cut into a "triangle" or any part of the frame that is still a solid structural shape. Many times after the first cut into a "triangle," such as cuts #1 and #5 in Figure 5-38, the space made by the saw or grinder will try to collapse, thus catching the blade and causing it to snag. If this happens while using your grinder, it could fly out of your hands if you aren't holding it tightly, so be aware. If this becomes a problem, just stop cutting with the grinder right before the end, then finish the cut with a hacksaw.

When you have finished cutting, you should have a nice pile of tubing and various bike bits on the bench (see Figure 5-41). The more old frames you can collect, the easier it will be to make whatever you want as your creative energy kicks into gear. Throw away any badly bent tubing, and organize the rest for later use.

Now we need to grind the leftover stubs from the bottom bracket and head tube. These are the most useful parts when making a custom frame of any kind.

Set the part into a solid vice with the stub facing upward and begin by grinding the end closest to you without digging into the good part.

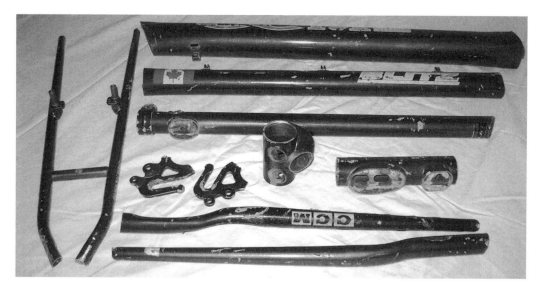

Figure 5-41   All parts of a frame after cutting.

For this job, use a proper grinding wheel, not the cut-off wheel, which is very thin and not made for side grinding (see Figure 5-42).

As you get the stub ground off, turn the work a little more and keep on going until all of the access tubing has been removed. Be careful not to take too much off or you will make the bracket or head tube too thin. When you are done, the piece won't look perfect, but it should still be fairly clean and round.

If you really want a clean job, grind until there is a small amount of stub left, then take it off with a hand file. Most of the time, this is not necessary because you will be welding another tube back onto the rough area soon enough anyway.

You will notice that most bottom brackets and head tubes have small holes where they meet the tube. This is normal. These are breathing holes that allow moisture to escape from the tube, preventing rust from forming inside.

If there are holes that seem to be as large as the frame tubes, or the entire frame looks like it were made from one giant tube, what you have is a "lug-type" frame, in which one tube actually fits into another. These frames were very common in older 10-speed type bikes.

With this type of frame, the bottom brackets and head tubes are no good because, after grinding, there will be large holes in them where the tubes were removed. If you have a frame like this, the only useful parts are the frame tubing and the rear dropouts, so avoid these

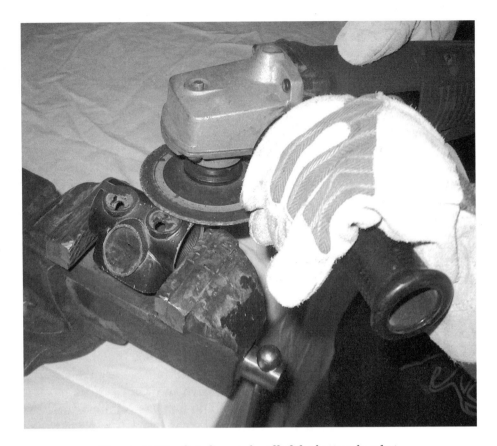

Figure 5-42   Grinding stubs off of the bottom bracket.

types if you can, or keep them for other projects that require an un-cut frame.

Since you have the grinder out, now is a good time to trim the bottom brackets as well. Put them in the vice and trim off the last bit of tubing left over from the cutting. Don't worry about the paint—this will burn right off during welding. Tubes and larger parts can be cleaned up with a wire wheel on your grinder as well.

When you are all done grinding, the parts will be clean of any stubs or welds and ready to be turned into whatever you like (Figure 5-43).

Now that you've learned how to get the "meat" from an old frame in this chapter, you should never refuse any bike parts in any condition because there will always be something for the hard-core bike builder to use.

Figure 5-43   These are the building blocks for many projects that you will create.

With a little imagination and a stockpile of ready-to-weld parts, you can make just about any type of custom bike, such as choppers (Chapter 6). Your initial attempts may not be completely successful, but at least you didn't have to search for three months to get all of the parts. You're ready to start building!

6

THE INFAMOUS CHOPPER

CONTENTS AT A GLANCE

Attitude Is Everything

Attitude—it's the one word that best fits the timeless and immortal chopper. Reaction to the chopper on the street will be mixed. Some people will give you a "thumbs up," some will turn away, and others will stop you to find out where they can get such a cool bike. For the rider, the chopper is a bold statement of independence and personal expression that challenges conformity. It's amazing what you can create from a chopped-up bike with extended forks. Build one and you will see what I mean.

The chopper is not a bike for everyone. If you want an easy-to-ride and comfortable bike, then this is definitely not the bike you want. Choppers by their very nature are awkward to drive, have a mind of their own, and put the rider in a less than ideal riding position between the long seat and tall handlebars. These are all ideal traits for a bike that refuses to be tamed and has an attitude similar to that of a wild steed.

But once mastered, the chopper and rider become merged into a single machine, and riding with grace becomes second nature. Once you are in control of your machine, it is always fun to let those you meet try to ride your chopper, and see them shake and wobble down the street as the bike defies their every attempt to gain control.

So, now that you know a little about the underlying nature of the chopper, you are ready to explore this chapter of extreme chopper bikes with forks boldly extended to ridiculous lengths.

Chopper Characteristics

What makes a chopper? This is a tough question to answer and is open for debate. Almost any radical change to the standard bike can be considered "chopping," but most people will agree that a true chopper is a bike in which the forks have been greatly extended. As soon as you make this modification, your machine will easily be identified as a chopper by the masses, and from here you may use your imagination to turn your machine into a work of art.

Extremely high and long handlebars, also know as "ape hangers," are also standard accessories for a chopper. These types of handlebars make the rider stretch out and hang on as if hanging off a monkey bar. You may ask why anyone would want to ride a bike like this, but the true bike rebels need no explanation—they just know it is just right.

In addition to the extended forks and ape hangers, the chopper usually needs a long seat, or "banana" seat. These types of seats were popular on kid's bikes in the 1970s and are making a comeback. The banana seat places the rider over the back wheel, creating a rear-heavy bike capable of pulling a wheelie at the slightest hint of acceleration, and adds to the difficulty of maneuvering and controlling chopper-style bikes.

With the three basic modifications—extended forks, ape hangers, and banana seat—any bike can be turned from a civilized ride into a crazy, rebellious machine. Any kind of ordinary bike can be made into a cool-looking chopper if you know the basic changes that need to be made. I have chosen such a bike, the "Dirty Rat," as one of the projects in this chapter (see Figure 6-1) just to prove my point.

Before you begin, you should have at least one set of ape hangers, a banana seat, and some spray paint for all of the projects presented

Figure 6-1   Choppers such as the Dirty Rat can be built in a matter of hours.

here. Ape hangers and banana seats are easy to find at yard sales and scrap yards, and can be purchased new at just about any bike store for a few dollars.

If you have a pipe bender and a lot of patience, you can even create your own wildly extreme handlebars for your chopper.

The intention of these projects is to challenge your creativity, so feel free to let your imagination run wild and build some outrageous choppers.

Now it's time to get your hacksaw ready and let the fun begin!

The Dirty Rat

A chopper can be made from just about anything with two wheels, or even three, and to prove this point, I have selected the most unlikely candidate for "choppification"—a girl's bike with purple flowers and hearts on the frame (see Figure 6-2). If I can turn something like that into a fierce, savage beast ready to take over the road, then you can, too.

For this project, you will only need an extra set of donor forks that are similar in size to the ones already on the bike so they can be force fit onto the ends of the existing to create longer forks. The longer the donor forks, the better. Some builders have even joined three or more sets together, but I would imagine that a machine like this would be almost impossible to ride due to the extreme curvature of the resulting forks.

Simple choppers like the Dirty Rat have been built by all ages since the invention of the bike and can be built in a single evening with only a hacksaw and hammer. Keep in mind that a little paint can make a big difference in the final product.

FEATURES TO LOOK FOR

When selecting a bike to "choppify," look for special features like banana seats, high handlebars, fat fenders, or odd-shaped frames. These features will help improve the look of your bike without a lot of extra work.

Now that you have selected a bike to chop, take inventory of what can stay and what must go. Tassels, "spokey dokeys" (round plastic objects on the spokes that makes sounds while the wheel is in motion), reflectors, pink tires, and stickers all must go, but, of course, you've probably already figured this out.

Figure 6-2 When choosing a bike to convert to a chopper, look for features such as banana seats, high handlebars, and fat fenders.

As shown in Figure 6-2, the bike I chose needs a lot of work to become a real chopper, but there are good things about it such as the fat rear fender, low frame, banana seat, and tall handlebars. You have to look past all of the accessories and envision what the final product will look like.

MAKING THE FORKS

The first and most important step in building a chopper is making the forks a ridiculous length to begin our transformation. Start by cutting off the legs of your donor forks as high up as you can.

Remember to select forks of similar shape to the original ones already on the bike so they can be put together (Figure 6-3). Most forks are either oval or round, and these two types will not be compatible unless you plan to weld them together.

Sometimes you will need to reshape the top ends of the freshly cut forks to fit them onto the ends of the original forks. Banging them with a hammer against a flat surface will work fine, but don't overdo it. Forks are quite thin and can be easily flattened with only a few good strikes of a hammer.

After the two parts are able to fit together an inch or two (see Figure 6-4), use a hammer to bang them as far in as they will go by striking the dropout end of the cut legs. It is a good idea to put a piece of wood between the hammer and fork end so you don't damage the dropouts to the point where the front wheel axle will no longer fit.

Also, try to get both legs an equal distance onto the existing forks so the wheel will not be leaning in one direction. Check this by fitting the front wheel on and looking down the end of the forks. If the wheel is not on straight, keep working the legs together until it is. Once the forks appear to be fairly straight and sturdy, it is a good idea to drill a small hole through one side and put in a screw or bolt (see Figure 6-5) as, over time, the legs will become loose.

You won't look very cool on your new chopper if the front forks fall apart while you are riding! Don't worry too much about the ugly seam

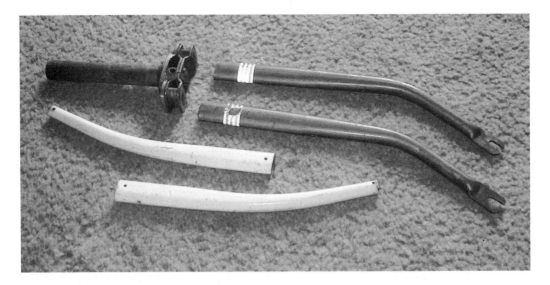

Figure 6-3 Select similar fork sets to fit together to create the elongated fork effect characteristic of choppers.

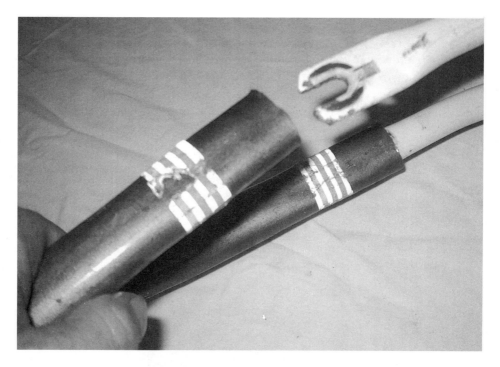

Figure 6-4   Attaching fork sets together.

Figure 6-5   Forks held in place with a bolt.

or bolt at the fork joint. This can be covered with a rubber tube or filled with body filler then painted later.

Before you drill the holes for the bolt, sit on the bike and make sure things are fairly straight. If you joined three or more fork legs, this may be quite a chore, but can be done with a little patience and a lot of hammering. The next important chopper modification is the handlebars.

MODIFYING HANDLEBARS

If the handlebars on your donor bike are just too small or low, it is best to find a good set of ape hangers, as a chopper is not complete without them. Many bikes have the crash-pad style BMX handlebars as shown in Figure 6-6 and, by cutting the cross bar, they can be converted into a decent set of chopper bars. Using a hacksaw, cut each side of the crossbar off (see Figure 6-6), then hand file the leftover metal.

Figure 6-6   Cut the crossbar off BMX-style bars to make chopper handlebars.

Now you have instant ape hangers (see Figure 6-7). If you want the handlebars higher up, you may want to look for a taller gooseneck. Most bikes will have a gooseneck that can be raised about four or five inches, but you can get one from a bike store that will give you 10 or more inches of height for that radical look. Sometimes, the gooseneck on an old exercise bike will also work, but these can be several feet tall.

You may also want to replace the handlebar grips if they are not your style. If they seem to be stuck, cut them off with a sharp knife rather than spending hours twisting away at them.

ADDING ACCESSORIES

Once you have the basic chopper taking form, all that's left is to add accessories and paint. Your bike now has long forks, a banana seat, and tall handlebars, but still lacks the attitude that a chopper should have (Figure 6-8), so a little more work needs to be done.

Figure 6-7 Ape hanger handlebars.

Figure 6-8 The Dirty Rat chopper is ready for accessories and painting.

Adding a nice fat rear tire and thin front tire, and choosing the right color of pedals, grips, and seat can make a big difference in the look of your bike. It's best to add all the little things and save the painting for last, so that you can avoid scratching your painted frame as you try different parts on the bike. Once you are satisfied with your creation, it may be a good time for a test ride to make sure it's actually rideable.

If all goes well, you should be able to beat the "shakes" after a few blocks and drive in a straight line with little effort. If things don't go very well and controlling the bike is a problem, take a section or two out of your 12-foot long forks to make it rideable. Once you overcome obstacles and make final adjustments, it's time to paint.

PAINTING THE FRAME

Cleaning the frame can be the hardest part of finishing your chopper, but it is worthwhile if you want a good finish. Painting over dirt and stickers is not recommended. Most of the stickers on frames can be peeled off with a little patience, as shown in Figure 6-9. Just get one of the corners going and slowly peel each strip until they are all gone.

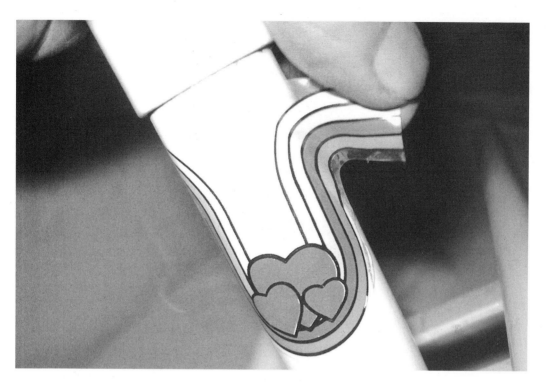

Figure 6-9 Remove all stickers and dirt before painting the frame.

There will be a tacky residue left over from the sticker that can be removed easily with paint thinner.

Soak a rag in thinner then wet the gooey area and let it sit for a minute or two. The sticky substance should rub off with little effort, but you may need to repeat this procedure several times to get it all off. If you are a perfectionist, you can sand the old paint off with an emery cloth but, for most paint jobs, this is not necessary. If you give the old paint a fine sanding just to rough it up a little and clean the dirt and rust off, it should take new paint with no problem. If the old paint is really rough, a good spray can of primer will also make a big difference in the final coat (see Figure 6-10).

One can of spray paint is usually enough to cover a bike, unless your forks are extremely long. Once your frame has dried, add all the parts back on, and polish any dirty chrome or bare metal parts. A small rubber tube or clamp can be used to hide the joint in the fork, or it can be filled before painting.

Accessories like chrome valve caps, a light, and cool-looking pedals can also help to spice up your creation after it is all back together. If

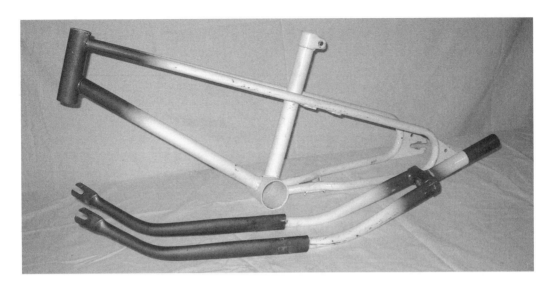

Figure 6-10   A clean frame, free of dirt and stickers, is ready to be painted.

Figure 6-11   The final product is ready for the road.

you are good with a small brush, pinstripes, flames, or designs on the frame will also give your machine that personalized look (Figure 6-11). Now it's time to cruise down the street and show off your work.

You may be surprised at how many other choppers come out of nowhere to join you on your ride. Riding in a group with other chopper riders is a lot of fun. Many cities already have several chopper bicycle clubs you can join, or you can start your own club or search the Internet to connect with people who share your interests.

Building choppers is a fun hobby that anyone can do. You may soon find yourself making something more radical or building your next frame from scratch. Check out the other chopper projects in this book if you want to take on more bike building creations of an extreme nature.

The Skyscraper

As shown in Figure 6-12, it should be obvious why this beast is called the "Skyscraper." With its radical fork length and extreme height, this

Figure 6-12   The Skyscraper.

is one extreme chopper. But building this machine only requires a few cuts with a hacksaw and two pipes to be welded to extend the forks.

FEATURES TO LOOK FOR

If you do not have a welder, you could have someone else weld the joints for you, and even if you had to pay for this service, it's a simple weld to make and wouldn't cost very much. For this project, you will need a ladies' roadbike frame (see Figure 6-13) and about 10 feet of one-inch electrical conduit or thin-walled steel pipe.

Look for a ladies' frame bike, so when we cut down the seat tube, it will lower the frame without changing the overall length. These frames are very common and easy to find just about anywhere. The main feature of this frame is that the main tube is actually made of two smaller pipes that continue to become seat stays. This type of frame makes a nice-looking tall chopper due to its straight and simple lines.

On some of these ladies' frames, the main tubes do not continue into the seat stays or they may only have a single main tube. These will be of no use, since we are going to be cutting the top half of the frame off. The frame should look very similar to the one in Figure 6-12.

Figure 6-13 A ladies' bike is an ideal frame with which to build the Skyscraper.

CUTTING THE SEAT TUBE

Once you have found a good frame candidate, get out your pipe cutter or hacksaw and begin by cutting the seat tube about two inches above the lower seat stays, as shown in Figure 6-14.

You will need the extra two inches to put the seat post clamp back over the tube to hold the seat post secure.

Once this tube has been cut, you will need to cut the two top seat stays off right at the rear dropouts. This will leave you with a long but short frame ready to be turned into a radical chopper with extreme attitude! Don't toss out the leftover frame parts just yet, as we will need to put the seat post clamp back onto the newly chopped frame.

The first thing that needs to be done before adding the seat post clamp is to cut the small slit out of the tube (see Figure 6-15). This can be done with a hacksaw or grinder. Look at the original tube to see the approximate width and length so you can get something close. This cutout allows the frame to squeeze against the seat post as the clamp is tightened and is necessary in order for the clamp to work.

Figure 6-14 Using a pipe cutter to cut the seat tube.

Figure 6-15   Cut a small slit in the seat tube.

Once you have the cutout made, remove the clamp from the cut part of the frame and place it back on the new part. Sometimes, the clamp is spot welded, so a little work with a hammer and chisel may be needed to get it to come apart.

CREATING THE FORKS

Once you have the frame cut down, it's time to create the best part of the bike—the super-long forks. For this, you will need a pair of mountain bike forks with a thickness of about one inch and a piece of pipe that is about the same length. This is a fairly common fork size. Electrical conduit is available in this size, so finding these parts should be easy.

Start by cutting the mountain bike forks somewhere near the top on the straight part, not the curved section, as shown in Figure 6-16. Try to get both cuts in the same spot so the two legs are the same length. Using a pipe cutter for this job will ensure an even cut all the way around as well.

Figure 6-16   Cut the bike forks below the curved section.

Once you have the legs of the forks cut off, stand up the leftover fork top on a level surface to see if they are straight, and if they are uneven, grind or file one of the legs until they are the same.

Now you need to decide how long to make the forks. Before you decide on 20 foot forks, remember that the longer the forks, the higher up this bike will get until it falls right over backwards. The forks I used are just less than 4 feet long, and at this length, the bike will do a wheelie even at the slightest lean backwards or fast acceleration.

If I made them any longer, riding the bike would be impossible without a load of bricks to weigh down the front end. If you want to make longer forks, you will need to cut the frame and lower the angle of the head tube, but this becomes quite a complicated task, so try to stay under four feet long.

Before you begin to join the forks, you need to remove the dropouts from the cut fork legs, as these will be put back on to the ends of the tubing to complete the forks. It is best to begin by welding the two dropouts to the end of the pipes (see Figure 6-17). This way, you can connect your front wheel to the tubes to aid you in welding the tubes to the top of the forks, ensuring proper length, width, and angle of the finished fork.

Weld the dropouts to the fork ends so they are straight and centered just as they were on the original forks. Once they are welded solid on both pipes, connect the front wheel so it will hold the two parts securely. Now for the tricky part.

ALIGNMENT IS IMPORTANT

With the top of the forks held in a vice, set the two new legs on the bench so they meet up with the top of the forks for welding. You may need someone to hold the front wheel, or prop it up with something so that it is straight while you tack weld the two legs.

Once the tack weld is holding the two legs to the top of the forks, make any adjustments you need to ensure that the front wheel is actually straight and centered in the forks. This operation may take a little time and patience, but is not that hard to do. When finished, you should have a nice straight long pair of forks for your chopper (see Figure 6-18).

Figure 6-17   Weld the dropouts to the end of the fork pipes.

Figure 6-18 Tack weld the legs and forks together.

When you're satisfied with the tack weld and alignment, remove the wheel and hold the forks up so you can inspect the job. Looking down the length, what you should see is a straight, even set of forks (see Figure 6-19), not something that looks like a pretzel. If there is a lot of distortion or if the legs are badly uneven, try again until you get it right. This step is crucial, and it will make or break the entire project.

Now is a good time to put on a seat, some wheels and put the forks on the bike to make sure they are not too long or badly out of alignment from side to side. Choose the same size wheels you plan to use in your final design when doing this check. In my designs, I like a smaller front wheel, so I chose a 24-inch rear wheel with 20-inch front wheel (see Figure 6-20).

When you have everything together, sit on the bike and look down the forks to see if the wheel is straight. If the wheel is leaning to one side or if you instantly fell back as the bike went into an uncontrollable wheelie, it may be time to redo the forks.

If all looks OK, and it seems like the bike will work properly, you can complete the welding and grinding of the forks. If the pipes you chose to extend the forks with were a little bigger or smaller than the original forks, then you may need to spend some extra time welding and grinding the highs and lows in order to make a smooth transition in the weld. The forks I used were about an eighth of an inch smaller

Figure 6-19 Make sure the forks and legs are straight so that the wheel is centered.

than the pipe, but after some careful welding and grinding, they turned out fairly smooth, as shown in Figure 6-21.

A good coat of primer and paint can go a long way in covering up imperfections and small grind marks in the steel.

ADDING ACCESSORIES

Now that you have the hardest part of this project completed, you can begin to add the standard chopper accessories such as banana seat, ape hangers, fat tires, and fenders to beef up your cruiser (see Figure 6-22). Although it's pretty hard for a bike with four-foot long forks to look boring, adding all the trimmings and a good paint job can make the bike look even better. Find some really wild handlebars, a banana seat and a rear fender, and now you have the look!

If you are having a hard time finding a fender, you could always salvage one of the bent, rusty ones that probably came with the donor bike. With a little effort and a hammer, the old bent and rusty fender

Figure 6-20 Choose same-size wheels or two different sizes to enhance the Skyscraper's look.

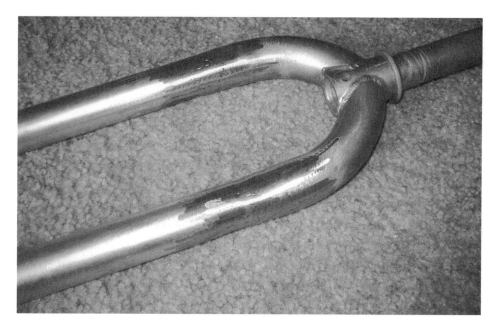

Figure 6-21 Grinding welds to achieve a finished look before applying primer.

Figure 6-22   The right accessories, such as a banana seat, can really enhance the look of the Skyscraper.

can be straightened and painted to look just like new (see Figure 6-23). A smaller hammer with a round head is good for taking dings out of the fender, but don't hit the thin metal too hard or you will just make it worse.

FINAL ADJUSTMENTS AND PAINTING

Once you have all the parts you want for your chopper, put it all together one more time just to make sure it's all going to fit, then take a test ride. This will be you final test before painting, and your last chance to make fork adjustments if needed (Figure 24).

Although your first few trips around the block may a little shaky, after a short time, you should be able to handle the bike in a smooth,

Figure 6-23 Bent, rusted fenders can be restored to look like new.

Figure 6-24 Make final adjustments before priming and painting.

straight line without a problem. You will notice that this bike is always ready to do a wheelie, so be careful while riding up a steep hill.

Once you are sure the bike is actually rideable for more than 10 feet at a time, and everything fits together properly, you can give it a fresh paint job, buff up the chrome parts, and add all your latest accessories.

When you take your first ride, be prepared to stop and answer a lot of questions. "Did you make that?" "Where can I get one?" "Will you make me a chopper?" "Is it for sale?"

As you've already seen in this chapter, choppers are fairly easy to make. The next project, the Highlander, is a bit more challenging, but the end result is worth the extra effort.

The Highlander

Without a doubt, the Highlander is a chopper that oozes attitude and commands respect. Based on the wild "easy rider" style low-riders of the early 1970s, this extreme bike is fun to ride and draws a lot of attention on the street from young and old alike. Although this project is custom built from the ground up, it requires only a few lengths of electrical conduit for the frame, a small piece of sheet metal for the fender, and some basic bicycle parts. This project can be built in a few evenings with only the basic tools and a little patience.

This bike is built to fit the rider and fork length is entirely up to you, so I will not be presenting a strict set of measurements for the design. These measurements will change according to your leg length and design style. You may want the bike to fit a range of rider sizes, or set a new world record for fork length. This is all possible, but there are some basic design principles—such as keeping the pedals high enough off the ground to avoid hitting the ground during a turn, and making sure your legs can actually reach the pedals—that need to be followed.

BASIC MEASUREMENTS

As shown in Figure 6-25, there are five basic measurements. Measurement "A" will always be the same, and "B," "C," and "D" are a product of "A" and "E." Confused? Sounds like we are going to build a space shuttle for NASA here, doesn't it? Well, have no fear, this is a lot simpler than it sounds.

Basically, length "A" is the same because it puts the rider a certain distance from the handlebars. If you are 6 feet tall or 4 feet tall, this

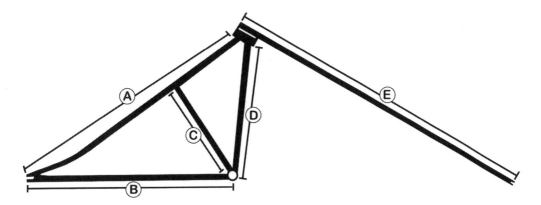

Figure 6-25 Basic measurements for designing the Highlander frame.

length will be just fine, since the handlebars are easily adjusted forward or backwards. It has also been designed so that the frame can take a 26-inch rear wheel with a fat tire.

Measurement "E" is, of course, fork length (the very essence of a chopper). This is something you will decide on, although I will make a few suggestions for practical riding purposes.

OK, now that we know the length of "A" (always the same) and the length of "E," (the forks), we can than calculate "B," "C," and "D" so that the rider can reach the pedals and the pedals do not scrape on the ground. That's all there is to it.

PARTS YOU WILL NEED

As always, before you start, it's a good idea to source out all the raw materials for the project. Since the biggest part of building this chopper is your own hard work, the list of materials and cost is very small, and most of the parts like those shown in Figure 6-26 can be salvaged from scrap bikes. Here's what you need:

- About 20 feet of one-inch, thin-walled steel electrical conduit (also called EMT).

- Two sets of forks from a 26-inch mountain bike (these do not have to be exactly the same, just as long as they are for a 26-inch wheel).

- One head tube and fork set salvaged from another bike frame (any size fork will do as long as the head tube and forks fit together).

- One bottom bracket and crankset salvaged from another bike frame.

Figure 6-26 Many parts used to build the Highlander can be salvaged from scrap bikes.

- One gooseneck and set of ape hanger handlebars.

- A sheet of 24-gauge or similar thickness steel for the fender (about 4 feet square).

- Two nuts and bolts for the forks (nuts need to be about 1 inch in diameter to fit inside the electrical conduit, as you will soon see).

- One big seat with springs (these can be found on many exercise bikes).

- One 26-inch rear wheel with coaster brake and one 20-inch front wheel.

- A few chains of equal size to join together (these must fit on your crankset).

Once you understand how all the parts go together, feel free to modify the design to suit your needs or style. You may want a 24-inch front wheel, or 20-inch on the front and rear, or even 36 speeds with front and rear shifters and disc brakes.

MAKING THE MAIN TUBE

The first step will be to cut the main tube "A." Since the length of this part of the bike includes a set of forks, you must put them together before you mark the tube for cutting. As shown in Figure 6-27, the fork threads are put inside the tube, and then measured from the end of the forks to the end of the tube to be cut. The bearing ring is no longer needed, so just tap it off with a hammer and add it to your parts collection for later use.

The total length from the fork tips to the cut on the pipe should be 44 inches. Once marked, you can cut the main tube with a pipe cutter or hacksaw, and also cut most of the head tube from the set of forks. This will not be used since it will be inside the main tube after welding.

Once you have the tube cut, weld the tube and forks together where they meet at the base. Grind and clean the weld before you move ahead, as the part is easier to handle at this stage since there is nothing to obstruct you.

MAKING THE FORKS

Now that you have the main tube welded to the forks, you must decide on how long to make your front forks, as these two parts of the bike will dictate how the rest is made. In my design, I used a total fork

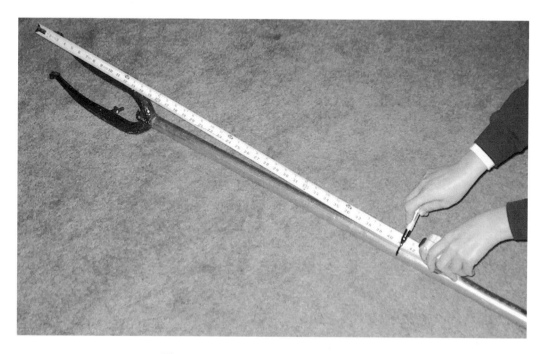

Figure 6-27 Measuring the main tube.

length of 56 inches (measured from the very top of the forks to the dropouts). You could probably make longer forks if you wanted to, but after making a few choppers, I found that this length is about the maximum length you can have on a bike and still have decent control.

This length also allows you to turn around in the width of an average city street without having to pick up the bike and turn it around manually. Of course, if show is your main reason for making this chopper, you could use the entire ten-foot length of conduit for the forks and still make a rideable bike, although it will be quite a challenge to ride!

Once you decide on a reasonable fork length, cut two identical lengths of conduit. It is important that both pieces be the same length, or your front wheel will be on crooked, and steering will be difficult.

Lay the newly welded main tube and one fork length on a flat surface, as shown in Figure 6-28. A long stick or any straight object is placed as shown in the figure so that you can visualize your bike standing up. When the basic frame is complete, the front and rear dropouts and bottom bracket should all be on the ground at the same level. This makes it easy to design the bike so that the pedals will not hit the ground. Set the forks so that they are roughly 90 degrees to the main tube.

MEASURING LEG LENGTH

You now need to figure out your leg length (inseam) so you can set the proper distance from the seat to the bottom bracket. If you are building this bike to fit several riders, measure the inseam of the

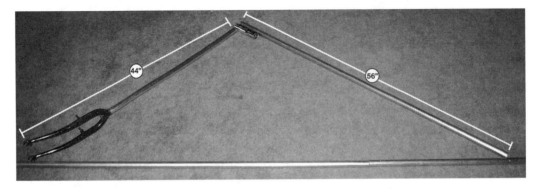

Figure 6-28 Lay the basic frame on a flat surface and place the forks 90 degrees to the main tube.

shortest rider, but don't expect a bike made for a four-foot-tall rider to fit a six-foot-tall rider. If there is more than five or six inches difference, you will need separate bikes or some method to adjust the seat height.

The easiest way to measure your inseam is to stand with your back against the wall and have someone place a ruler between your legs and measure the distance from the ruler to the ground, as in Figure 6-29. You can also set up a "regular" bike (remember those?) so that it is adjusted for your height, and just measure the distance from the top center of the seat to the center of the lowest pedal. Once you have this measurement, mark it down for future reference because it will be important for this project and many of the other ones in this book.

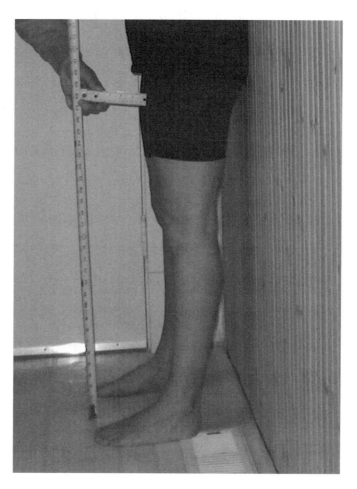

Figure 6-29 Measuring a rider's inseam.

MEASURING SEAT-TO-PEDAL DISTANCE

OK, now for the most critical measurement—the distance from the top of the seat to the farthest position of the pedals during a full crank rotation. To set this distance, put a tape measure in the area in which the seat will be (about four inches past the joint where the main tube meets the forks) and pull the tape 90 degrees away from this point until you have measured the distance of your inseam (see Figure 6-30).

Place a crank arm on the guide that is being used to line up the frame so that the end of the arm is as far away from the seat area as possible. This will be your endpoint. You will also notice that I added some extra distance by starting the measurement about three inches away from the main tube as well.

This extra distance compensates for the extra thickness the seat will add. If you imagine yourself sitting on the seat pedaling the bike, you'll notice that this distance (your inseam) will be the farthest distance your legs will have to travel during a full pedal rotation. If you used a regular bicycle to take this measurement, it should be the same.

MEASURING FORKS

Now place the second set of forks alongside the guide so the tips meet the other set of forks, and point the threaded end toward the crankset

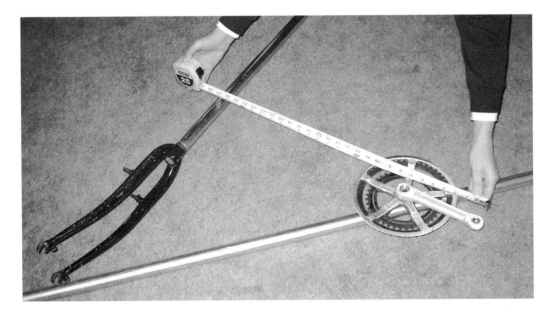

Figure 6-30 Accurate seat-to-pedal measurements are crucial to the Highlander design.

(see Figure 6-31). These two forks will be welded together to create the rear end of the frame. You can use curved or straight forks depending on the style of frame you want, but if one set is heavier than the other, try to use the heavier set on the bottom.

Measure the distance from the center of the crankset to the joint in the forks where the tube is welded to the top, and cut a tube to that length. You will be placing this tube between the bottom fork and the bottom bracket.

If your forks still have the bearing rings attached, remove them so that the threaded part can slide all the way inside the pipes you are using for your frame.

MEASURING THE LOWER FRAME

Once you have the pipe cut for the lower part of the frame, drop it in place, and place your bottom bracket (salvaged from another frame) onto the end of the pipe, as shown in Figure 6-32. Remember that this pipe is parallel to the ground, and draws a line from the rear dropouts to the front dropouts. If you imagine the completed frame standing upright on level ground, the rear dropouts, bottom bracket, and front dropouts would all be touching the ground at the same time.

Measure the distance from the bottom bracket to somewhere in the middle of the top tube, and cut another pipe to fit (see Figure 6-33). The actual angle and position of this tube is not critical, so feel free to

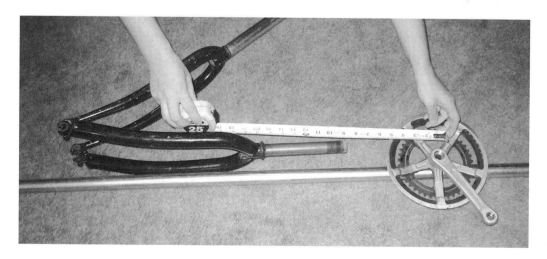

Figure 6-31   Measuring the forks to create the frame rear.

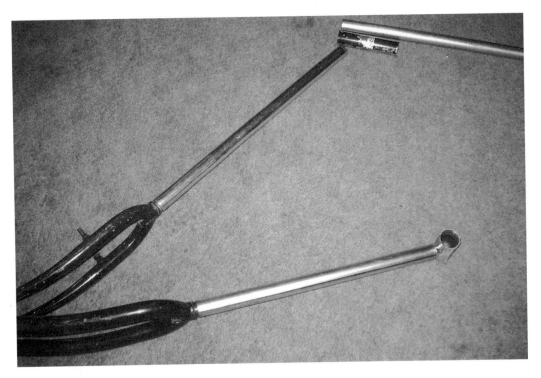

Figure 6-32   Joining the bottom bracket to the lower set of rear forks.

place it to suit your taste. In my design, I found that a 90-degree angle from the top tube to this one looked good, but you can bring it back right under the seat if you want to. Don't worry about the exact lengths of your cut tubes just yet, as we will be grinding them all to fit together before welding anyway, to get a closer joint.

MEASURING THE FINAL PIPE

Now measure and cut the final pipe that will connect the bottom bracket to the head tube that has been salvaged from another frame. At this point, you can see the frame take its basic form (Figure 6-34), and it is a good idea to check your inseam measurements again, just in case something is wrong. Remember, you will be sitting on a 2- or 3-inch thick seat just above the top rear set of forks and your leg will have to reach the farthest position of a crank rotation, so check it again! If you are satisfied that all looks good, then it's time to make some sparks.

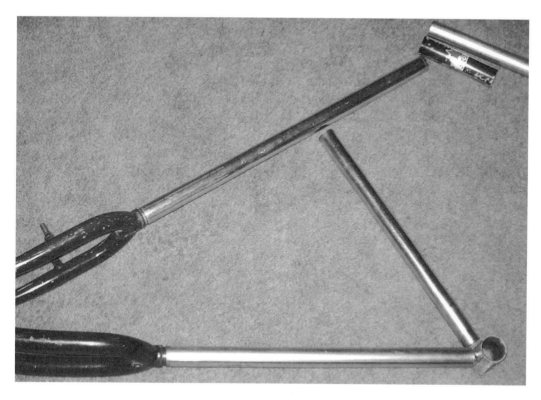

Figure 6-33 Adding more frame tubes.

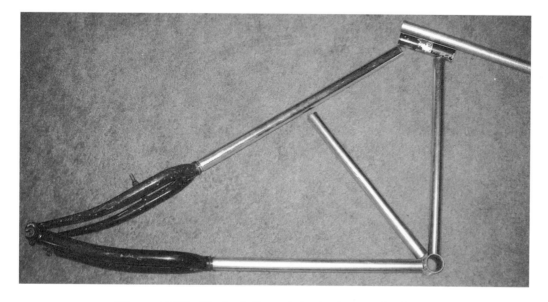

Figure 6-34 The basic frame is beginning to take shape.

WELDING THE REAR PARTS

The first weld will be at the very back of the bike where the two sets of forks meet. Cut most of the dropouts from the top set, leaving only a little bit of metal to weld to the bottom set, as shown in Figure 6-35.

Weld the two forks together at roughly the same angle they were at when you first set the frame up. It's a good idea to only tack weld these at first, so they can be adjusted as the frame goes together. Notice the heavy dropouts already on the lower set of forks. This is something to look for when choosing the donor forks. If the dropouts on your forks are very small or damaged, you may have to find a better set from another frame.

Once the two forks are joined, you can weld the other two tubes that fit over the threaded end. Since the threaded ends have no real use inside the tube, they can be cut before welding, leaving only an inch or two to help line up the tubes. Both of these welds will be final welds, so feel free to do a nice job and clean the welds while filling all gaps and holes (see Figure 6-36).

Once you have the welding done, look down the end of the two sets of forks, making sure the tubes are aligned with each other. If one tube is

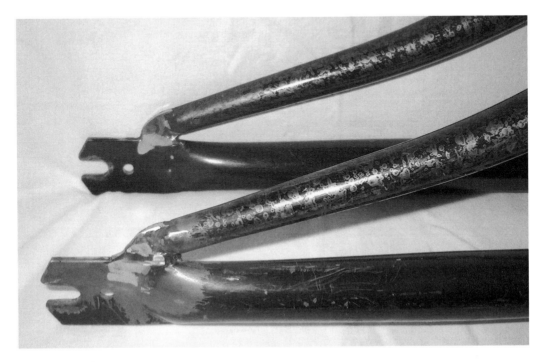

Figure 6-35 Leave some metal on the dropouts to weld to the bottom set.

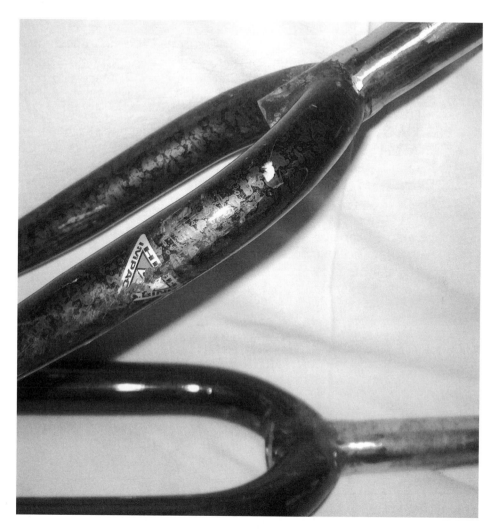

Figure 6-36 Clean welds by filling gaps and holes.

out of alignment with the other, you may be able to use a little force to get it straight. If the alignment is really bad, you will have to break one tack weld from the rear fork tips and fix it. At this stage, the rest of the frame will follow, so if you do a good job here, the entire bike will be straight.

WELDING THE REMAINING PARTS

Now you can begin welding the rest of the frame together using the same process you used to lay it out on the floor. Weld the bottom bracket to the tube that joins the lower rear forks, making sure that it's positioned at a perfect 90 degrees (Figure 6-37). It is important that the bottom bracket is straight so that the crankset will not be on an angle, because if it is, the chain may keep falling off. Once you have made sure that the bottom bracket is aligned correctly, finish and clean the weld.

Next, weld the tube from the bottom bracket to the head tube, and weld the head tube to the top tube using only tack welds. Your frame is now in the shape of a triangle and is only missing one tube—the one that divides the frame into two parts.

Check that everything is in alignment by looking at the frame from all angles, especially from the front and rear. The head tube needs to be aligned correctly or your forks will be on an angle. This angle will be quite noticeable because the forks are so long.

If something is wrong, tap the piece straight if you can, or break the weld and try again. Once you have it all straight, weld in the last

Figure 6-37   Bottom bracket positioned at a 90-degree angle.

tube—the one that connects the bottom bracket to somewhere in the top tube. This tube is left for last, since it can be placed just about anywhere, and the actual frame geometry dictates its length.

Remember that professional frame builders have jigs that hold all the parts together, but what you are doing is true art. Have patience!

THE GUSSET

After hours of welding, grinding, breaking, and rewelding, you now have the perfect chopper frame, right? Yes! The next part, the gusset, not only adds strength to the front of the frame, it enhances the Highlander's looks as well.

The gusset is made from a piece of sheet steel of approximately 12 gauge and can be just about any shape you like, as shown in Figure 6-38. I made a mock-up gusset out of cardboard just to make sure

Figure 6-38   The gusset is made of sheet steel.

it would fit, then traced it onto the sheet and cut it out using a grinder.

Once you have a piece that you like (and one that fits), weld it in place. I decided not to make a gusset that would go right up to the head tube because it would be hard to get in there with a grinder afterward to clean up the welds. But maybe you have better tools than I do, so feel free to experiment.

It's time to take a break so you can sit back and admire the work of art you have created. Your frame should now look something like the one in Figure 6-39 if all went well. Clean up all the welds with a grinder, filling in any holes and gaps before you move on to the next stage.

MAKING EXTENDED FORKS

It's now time to give life to your chopper and create the extended forks. Place your frame on the ground and hold one of the tubes that you will

Figure 6-39 Basic Highlander frame with gusset.

be using as a fork leg in the general position it will be in when the bike is complete. The fork leg should line up on the same angle as the head tube and be able to touch the ground in front of the bike (refer back to Figure 6-25).

Once you have checked this alignment, weld a nut into the end of each of the fork tubes. Try to find a nut that will fit snugly into the tube so it will align with the center easily. Weld the outer edge of the nut to the tube, being careful not to weld in the threads or over the hole (see Figure 6-40). If you think you may slip and weld into the threads, place a bolt or some other object into the nut first to protect the threads.

When you have both nuts welded into the top of the fork tubes, weld a set of dropouts (salvaged from another set of forks) onto the ends of each fork leg. You will notice that on my forks I welded an entire section of the donor forks onto the ends of the tubes, as shown in Figure 6-41. I did this because I liked the tapered look of the dropouts on the donor forks. This system will work fine, but you will have to find forks that are roughly the same thickness as the tubing, or there will be an ugly seam. Also, remember to cut the extra length from the new fork

Figure 6-40 Welding the nuts into the top of the fork legs.

Figure 6-41   Welded fork tubes and legs.

legs to compensate for the length of the donor parts, or you will be
making a Skycycle, not a chopper. No matter how you make the forks,
they should be no longer or shorter than your original plan.

TRIPLE TREE FORK PLATE

The forks you will be creating are of the "triple tree" type. There is an
upper and lower plate connecting the fork legs to the head tube. This is
the same style fork you will find on all motorcycles and even on some
beefy downhill racers, due to its extreme strength. This style was chosen for both strength and looks. You will need a plate of 1/8-thick steel
with enough room to draw two six-by-three-inch squares, as shown in
Figure 6-42.

If you look ahead a bit further in this chapter, you will get a better
idea how this fork set is going to fit together. Basically, the lower plate
is welded to both fork legs and the threaded tube, and the top plate
bolts onto the fork tops and head tube.

Figure 6-42   Draw the triple tree fork plate pattern on a plate of steel.

Start by drawing out the two six-by-three-inch boxes on your plate. After you have the two boxes drawn on the sheet, make straight lines from the lower point of "D" to the corresponding point at "C" (see Figure 6-43). Measurement "D" will be one inch in length, and "C" will be two inches, or one inch from each side of center. The lines will then create the basic shape of the tree plates, and they can be cut from the plate with a torch or grinder.

To create the proper placement of the holes for the fork legs, place a piece of tubing with the same diameter as your fork legs into each of the corners of the plates as far to the edges as it will go without hanging over and trace a line around it.

For the lower plate, this traced area is ground or cut out so the fork legs will fit into the hole for welding. For the top plate, you need to find the center of each circle you traced and drill a hole large enough for your bolt (the one that fits the nut welded into each fork leg) to fit through. The top plate also has a hole for the fork's threaded tube to fit through. This hole will be slightly larger than one inch so it fits snugly around the threaded tube, but not so tight that it damages the threads.

Once you have the holes cut into the plates, you can round off the edges with a grinder, as shown in Figure 6-44. Make sure that you leave enough area around the top plate to cover the fork legs once they are bolted in place. When grinding the edges, place both plates together in a vise, so that they will be of equal size when you are done.

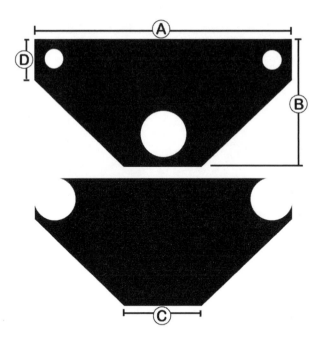

Figure 6-43 Measurements of the triple tree plate.

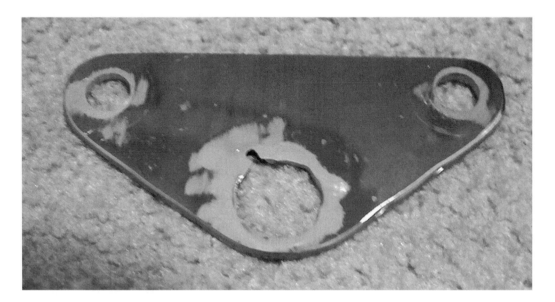

Figure 6-44 Use a grinder to round edges of the triple tree plate.

WELDING THE THREADED FORK TUBE

The next step is to weld the threaded fork tube (salvaged from another set of forks) to the bottom tree plate. The threaded tube is cut from a pair of old forks so that there is about one inch of metal remaining under the bearing ring (Figure 6-45). Also, make sure that this threaded tube is a good match with the head tube that you welded onto your frame, as they need to fit together properly. The correct size will allow you to put the bearings in the cups and get all of the caps on the top, just like they were on the donor bike. Check this before making any welds.

Once you are certain that the threaded tube is the correct length, place the top tree plate over the bottom plate and trace a line around the inside of the large hole in the top plate onto the bottom plate. This is where you will be welding the bottom of the threaded tube to the

Figure 6-45 Weld the threaded fork tube to the bottom of the tree plate.

bottom plate as shown in Figure 6-46. Make sure the tube is centered, and at 90 degrees with the plate. Referring to Figure 6-46, notice how the final fork set will fit together.

Once you have the bottom tree plate welded to the threaded tube, you must figure out at what distance to weld each fork leg onto the cutouts in the bottom tree plate. To do this, assemble the treaded tube onto the head tube, using all the bearings and the top ring (the one that rides into the top bearing cup).

Next, place the top tree plate over the top ring, and put the top nut on as it will be when the bike is complete. Now you can see exactly where to weld the fork legs to the bottom plate as this is how it all must fit together in the end.

Tighten the two bolts that hold the top tree plate onto the fork legs, and make a few good tack welds to hold the fork legs to the bottom plate.

Figure 6-46   The fork set and tree plate held in place with bolts.

To ensure that everything is straight before the final welds are done on the bottom plate, take the forks apart again, put in a front wheel, and look down the ends of the forks, as shown in Figure 6-47. If things are a little out of alignment, you may need to have someone hold the front wheel in place as you use a little force to align everything.

Although you could just adjust the handlebar position to compensate for a twisted set of forks, it will make your project a lot nicer looking if

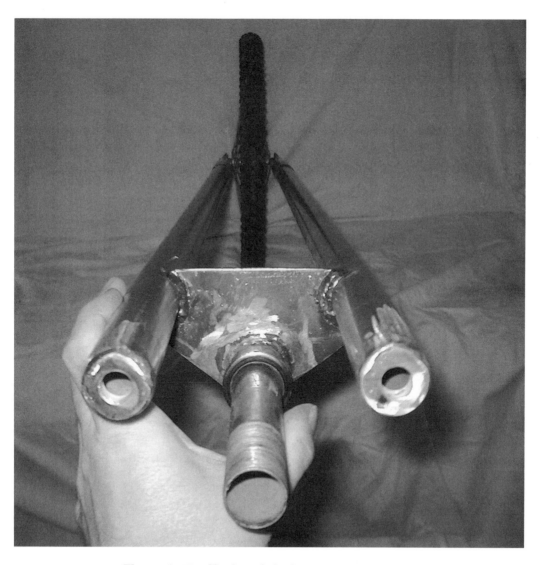

Figure 6-47 Checking fork alignment after welding.

it is just done properly in the beginning, so don't feel bad if you have to break welds and try again.

Once it is all looking good, finalize the welds to the bottom tree plate and grind the welds on the top fork tubes where the nuts are welded in.

MAKING A LOW-RIDER FENDER

Now for the "eye candy"—the big "phat" (cool) low-rider-style fender. Sure, you could just slap on a simple chrome fender from an old road bike, but why skimp now, after all this work? This part of the bike involves a lot of welding and grinding, but it's not complicated.

You will need a sheet of 12-gauge or similar thickness sheet steel that has enough room to draw two 28-inch diameter half circles on it (see Figure 6-48). The basic idea is to create two sides and a top, then weld it all together. Sounds easy enough, right? First, get whatever rear wheel you plan to use on the bike, fit a tire onto it and inflate it. If

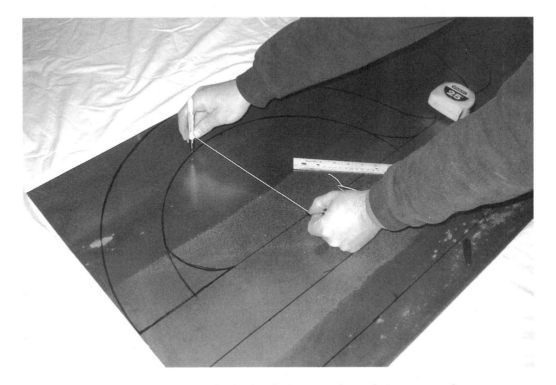

Figure 6-48   Draw the fender design on a sheet of 12-gauge steel.

you plan to use a nice fat rear tire, put it on now because we need to know what diameter to make the inside of the fender.

Take a measurement from the axle of your rear wheel to the farthest edge of the tire and add 1.5 inches to this distance. This will allow for clearance.

Tie a string to a marker, place it on the sheet, and draw a half-circle so that the edges of the arc are at the edges of the sheet. This minimizes waste. Now decide how deep you want your fender to be, and draw the inside arc from the same point. I made my fender four inches deep, but you could make yours less or more, or even tapered if you like.

You will need two sides and a top for the complete fender (see Figure 6-49). The top strip is as wide as the distance between the top set of fork legs that make up the rear of the bike, as you will be welding it there. My fender is 2.5 inches wide; this is a fairly standard width for a pair of forks. The top strip also needs to be as long as the distance from the front to the back of the fender, so you will have to make this meas-

Figure 6-49 Cutout pieces of fender ready for welding.

urement with a flexible tape. You will also notice that I added some styling to the front and rear of the fender sides, so make sure the top strip will fit over the entire length.

Making and Welding the Fender

Once you have drawn the fender template, cut it out using whatever method works best for you. I used a jigsaw with a metal cutting blade. Although it was a long process, the final product needed almost no grinding or reshaping before welding.

Welding the fender together is an easy process. Lay one of the sides on the bench and begin to tack weld the center strip at one of the ends. Bend the flexible strip along the side as you place another tack weld about two inches from the first, and repeat this process until the entire strip is tack welded to the side.

Then place the other side on to the strip and tack weld it in the same manner. The completed tack welded fender will look like the one in Figure 6-50. Now, check the fender for straightness and adjust it so that both sides are 90 degrees to the top strip.

When your fender is set up straight, weld the entire distance on both sides from the top, then grind and fill any holes. This as a long process, but will result in a very nice-looking final product, as shown in Figure 6-51. You may also want to use a flexible grinder disc or power sander to really smooth out the joint. Now take a break and collect your wits, because the next step will be to join the fender to the frame, and this is a job that may take some patience.

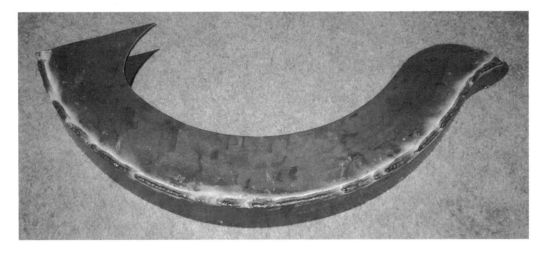

Figure 6-50 A completed tack welded fender.

Figure 6-51   A fender after grinding, ready to be joined to the frame.

As shown in Figure 6-52, the fender is directly welded to the inside of the top rear fork. Although this is not a hard weld to make, getting the fender to line up so that the rear wheel is not rubbing inside is quite a chore. The first thing you need to do is place your rear wheel onto the bike with your chosen tire inflated.

Next, you will need to tape some 1.5-inch foam or wood blocks to the top half of the tire to hold the fender in the correct position. Once the blocks are in place, put the fender over the wheel and into the correct place on the frame. You also need to put wedges along the sides of the wheel so that the fender has equal spacing on each side of the tire.

Once you have everything blocked in place, make a few solid tack welds to join the fender to each side of the frame until it is fairly solid. Don't go crazy with the welding here, and allow plenty of time for the fender to cool, because your tire is only a quarter inch from the hot steel.

When everything has cooled, remove the blocking and wedges, and see if the tire rubs. You may get lucky right away, or you may spend the next hour bending the fender until you get it right but, with patience, it will all work out.

Figure 6-52 The fender welded inside the top rear fork.

When the tire seems to be moving freely inside the fender, remove it and complete the welds along the forks, taking care to not burn through or overheat the thin steel. When you put the wheel back in, you may need to repeat your adjustment routine because the heat may have warped or pulled the fender out of alignment again. Don't worry, it will fit if you just have patience. I had to take the wheel on and off my bike about 10 times before I found the right spot!

ADDING THE SEAT

If you managed to get past the fender alignment puzzle, then you can take a breath, because the rest is smooth sailing. You need a place to sit on the bike, so find yourself a seat. I chose a big, wide seat from an exercise bike for two reasons. First, it suits the bike, and, second, because the assembly that holds the springs to the seat could be removed and welded right to the frame, as shown in Figure 6-53. This allows the seat to be as low as possible, and even adds a little suspension at the same time.

These types of seats are easy to find on both exercise equipment and "granny" bikes (you know, the old blue-and-white-style seats). If you can't find a seat like this, anything will work, but just remember your inseam measurement, and try to keep the seat as low as possible to the frame. The seat will be placed on the frame so that it is as close to the fender as possible, so you get a nice low-rider-style chopper.

FINE TUNING

Now for the fun part—the prepainting test ride. Put on a crankset, handlebars, chain, and the rest of the parts, and see how it rides (see

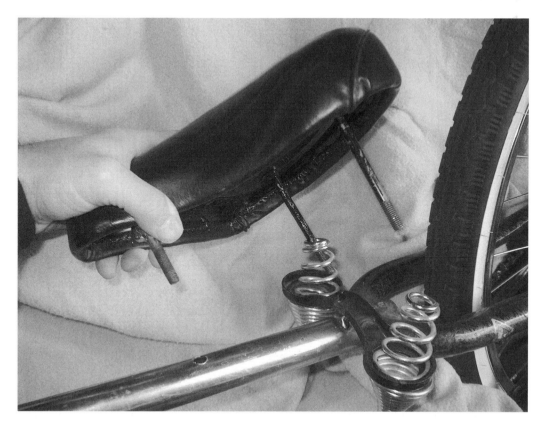

Figure 6-53 Some exercise and older-style bikes have springs that can be welded to the frame.

Figure 6-54). Does your fender rub? Are the forks straight? Do the pedals have adequate ground clearance? Does the chain stay on? Hopefully, everything turned out well but, if not, it's not too late to get out the hammer and make things straight. Chances are, your bike will ride perfectly and, after a block or two, it will feel just right. You will notice that these long choppers like to get moving a bit before they are totally stable, but once moving, they ride just as smoothly as any cruiser. Do you like all the "thumbs up" and horn blasts from your impressed observers? Yep, you know it—this bike is phat!

PRIMING AND PAINTING

OK, get off the bike for a while and strip it all back down to the frame and forks, because a bike of this quality deserves a nice coat of paint (see Figure 6-55). Use a good metal primer, and apply the paint accord-

Figure 6-54 Test drive your completed Highlander and make adjustments.

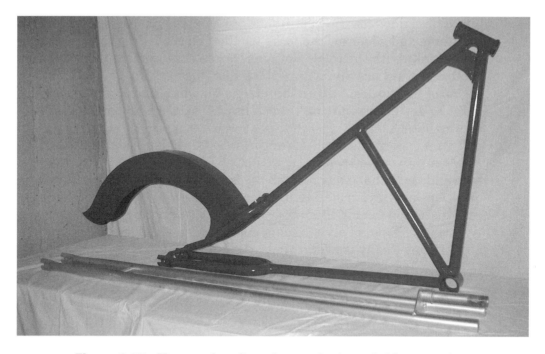

Figure 6-55   Use a good-quality primer and paint to finish your project.

Figure 6-56   Brad on his phat Highlander.

ing to directions for the best finish you can get. I chose cherry red for the frame and silver for the forks for the "old school" look, but feel free to get funky with the color. Hot pink, florescent green, multicolor paint splatters—they are all cool on a bike like this!

One of the hardest parts of building a bike like this is waiting for the paint to dry before putting it back on the road! Have patience, it will be worth it.

It sure feels good to roll down the street on this machine, doesn't it? How far do you get before crowds of people stop you to look over the bike and beg to have a test ride? Be prepared to stop often and answer questions about your design and techniques. Whether you build choppers discussed in this chapter or create your own designs, the chopper is always a popular style among custom bike builders (see Figure 6-56). Attitude really is everything.

WINTER WARRIORS

The Hammerhead

HAMMERHEAD CHARACTERISTICS

This chapter is designed to inspire you to build a bike to conquer the snow, ice, and slush of a typical winter climate. The Hammerhead is a two-headed monster that eats snow and ice for breakfast, and has no fear of Old Man Winter or his frozen wrath. I designed this machine to give myself a way of staying in shape during the winter months. I used to take my fancy, overpriced mountain bike out for winter rides, but I soon realized that it wasn't suitable in deep snow, or around icy corners, and the bike was taking a lot of abuse every time I bit the dust (I mean snow).

Obviously, a three-wheeled bike was necessary to maintain balance, so I rebuilt one of those old-style trikes (the kind with two wheels and a big basket in back) and tried to make it as light as possible by removing all parts that weren't needed, then added some knobby tires and went for a ride. The results were very disappointing. Not only was this bike still as heavy as a tank, but it also had no traction at all. That style of trike only drives one of the rear wheels, which mainly just spun around on most surfaces except bare pavement. Adding a differential (a gear system to spin both rear wheels and transfer power between them) was just too complicated and would add even more weight, so I decided to scrap this type of approach.

I would need two wheels up front instead. This design is popular on low-slung recumbent trikes that you find through the Internet (type "tadpole trike" into a Web search engine). They are very fast and comfortable, but not a suitable design for a winter bike for several reasons.

First, you don't want to be slung two inches from the slushy ground while riding in winter because you will get very wet. Second, people in cars will not be expecting many bikes that time of year, so you want to be as visible as possible. A low recumbent trike is not very visible to drivers of motorized vehicles.

The third reason is road salt. If you live in a community that routinely uses salt on roads and sidewalks, this is a problem because salt will corrode metal. Why spend so much time and money on something that will require many custom-made parts if it will end up rusted at the end of the year?

The Hammerhead is not only as high as a regular bike, but it needs only regular bike parts and a little welding here and there. The design uses a regular mountain bike with two head tubes welded on each side to allow for two front wheels. Both wheels steer at the same time just like skis on a snowmobile. In fact, the steering linkage *is* from a snowmobile!

The bike is called Hammerhead because I thought the finished frame looked something like a hammerhead shark. You see that, don't you?

PARTS YOU WILL NEED

Now that you have a way to conquer winter, let's start by gathering some parts. As shown in Figure 7-1, you will need a complete mountain bike (stripped down to the frame), two front wheels, two head tubes (ground clean), and matching forks. The critical requirement here is that both head tubes, forks, and front wheels be identical in size. Even the tires should be the same, as any mismatch will cause the finished bike to be uneven and wobbly.

CREATING THE TWO-HEAD TUBE EXTENSIONS

The first step is to create the two-head tube extensions. Each head tube is welded to two, 12-inch lengths of one-inch, thin-walled electrical conduit or similar tubing. These two pipes are then welded to each side of the original bike's (main frame) head tube. Both tubes are welded at exactly 90 degrees to the head tube, as shown in Figure 7-2.

It doesn't matter how far apart the two pipes on the head tube are, as long as they are not too far apart to connect to the main frame's head tube. If your main frame's head tube is a little shorter, take this into consideration when making these parts. You will also notice that I ground

Figure 7-1 Gathering parts for the Hammerhead project.

Figure 7-2   Creating the two head tube extensions.

the end of the tubes to mate with the head tube, forming a cleaner joint. This makes welding easier. Weld carefully, tacking it only at first to ensure that the tubes end up at 90 degrees to the head tube. Any error here will result with a front wheel on an angle. Feel free to browse ahead in this chapter to see what the final frame should look like.

When you have both head tubes welded to their two 12-inch tubes, it's time to weld them to the main frame head tube. As shown in Figure 7-3, this is also done at 90 degrees. You want each head tube to end up at the exact same angle as the middle head tube, so take a lot of time to get it right here, as this is the make it or break it part of the game. If you imagine two identical bikes standing side by side, then you can picture what we want here.

Use a square to check all the angles, and weld it fairly strong, but not all the way around yet, just in case it's not as straight as it looks (we may need to use a little force for final alignment).

CHECKING WHEEL ALIGNMENT

To ensure that the two head tubes are accurately aligned, put in the bearings, rings, and forks on both sides, then connect the two front wheels. Remember that both front wheels must be identical, including the tire pressure, or something will be out of balance.

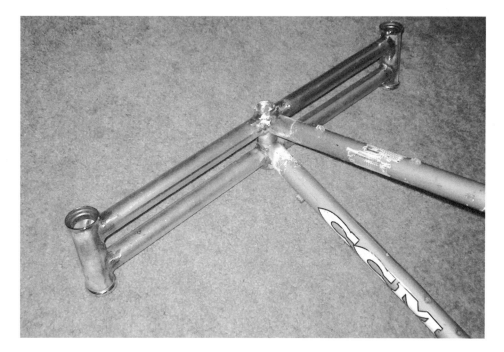

Figure 7-3 Weld the head tubes at a 90-degree angle.

When you have both wheels on, stand up the bike (see Figure 7-4) and really give it a thorough checking. Set the front wheels as straight as you can, taking measurements from each end of the rims or fork tips. Make any adjustments with a rubber mallet, and if your alignment is way out of whack, well, you know the routine—back to the shop.

Once you are certain that everything is straight, make your final welds, checking often to make sure heat distortion is not pulling things out of whack. Remember, this is important, so take your time and do it right. Another trick to getting this all straight is to take off the two front wheels and bolt or tack weld a threaded rod through both pairs of forks while making the final welds.

CROSS BRACING

The only other parts to add to this frame are the cross braces on each side. We have a lot of strength up and down because the four tubes hold the head tubes to the main frame, but there's not much strength front to back.

With a simple cross brace, we form a triangle on each side, making the frame extremely strong. Although I used the same one-inch tubing

Figure 7-4 Checking alignment of the two-head tubes.

for everything, you could use thinner tubing (see Figure 7-5) because
we only need compressive strength here.

This brace is welded from the top of each head tube somewhere near
the middle of the top tube on the main frame. The best place to weld
this tube is two inches in front of where your knee reaches while you
are pedaling the bike. To find this spot, put on a crank arm and set the
seat to your height, then mark it on the frame while you pedal. The
main goal is to make sure your knee does not hit the tube.

As shown in Figure 7-6, the finished, ground frame is a very sturdy
construction, and it will be able to stand a lot of strain. I tested this by
accident one time when I was out in a blizzard with very low visibility.
I was zipping down the sidewalk over a fresh layer of nice new powder,
and I had to veer off the sidewalk as I passed a few pedestrians. Because I could barely see and was traveling at quite a high speed, I hit a
fire hydrant and was launched over the handlebars onto my back in
the snow as I stopped instantly. Besides a face full of snow and total

Figure 7-5 Cross bracing on each head tube makes the frame very strong.

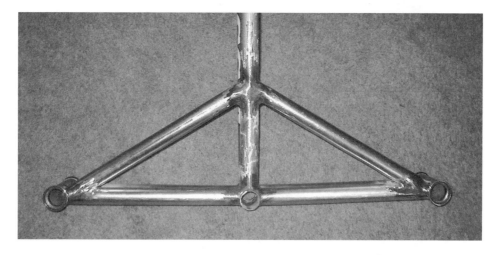

Figure 7-6 Finished frame with cross braces installed.

embarrassment, I was unharmed, and the bike only had a small scratch in the paint. Since then, I have ridden it down stairs and a riverbank, and done countless spinouts by locking up the brakes around corners, and the bike is still in perfect working order.

STEERING PARTS

Now that you have your basic frame completed, it's time to add all the steering parts. Take the original forks from the mountain bike, or a set that fits into the main head tube, and cut off both legs, leaving only the stem below the bearing ring, as shown in Figure 7-7. Once both legs are cut, add them to your giant bike-scrap pile for a future project, then grind the stub clean. This fork will have the handlebars mounted to it—this is why it must fit on the main frame's head tube. The plate that will connect both tie rods to the left and right forks will also be welded to the lower part of the stub.

To make both wheels turn at the same time from a central set of handlebars, we must fabricate a steering system similar to that of any vehicle with two front wheels.

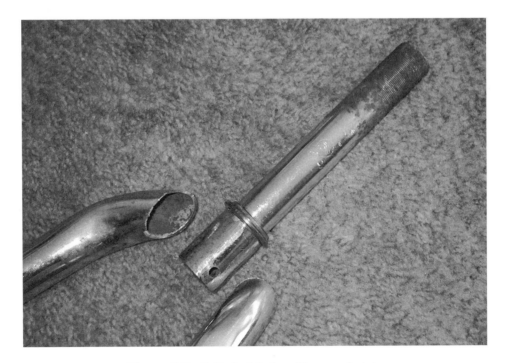

Figure 7-7 Delimbed forks with ground stub.

You will need a few inches of 1/8-thick steel plate or flat bar, four ball joints, and two thin rods about 12 inches long (see Figure 7-8). The ball joints can be salvaged from a trashed snowmobile, which is where I got mine, or bought new from a recreation or auto store. The heads on these ball joints are about one inch, but you could use ones half that size. Any performance auto dealer should also have something like this in stock. The two 12-inch rods will have the ball joints connected at each end, and can be any size in the area of ½ inch or so. The rods I am using are the two halves of the seat stay cut from an old frame. If you have access to a snowmobile, pop the cover and have a look at how the steering forks, as this will give you a better idea of what I am about to explain.

If you look ahead at the pictures in this chapter, you will see that both front forks are connected to the ball joints that are in turn connected to the rods and then to the main steering tube, so when you turn the handlebars, both wheels turn. What you might not have noticed is the slight angle that the plates that hold each ball joint to the forks are set at. This is called "Ackerman steering," and it makes one

Figure 7-8   The parts that make up the steering system.

side turn at a different rate than the other. You may wonder why you would want this. Let me explain.

Let's assume that you are driving around in a circle, turning to the right. The wheel on the right side (inside) is actually traveling less distance than the one on the left (outside) since the smaller a circle is, the shorter the distance around it gets. Because the inside wheel is traveling less distance in a smaller circle, it has to make a sharper corner, so it needs to turn more to the right than the outer wheel. If both wheels turned at the same rate, one tire would have to drag when you turn, and this would slow you down and ruin your traction. Understand?

Making the inside wheel turn sharper is done simply by setting the plates that hold the ball joints to the forks on an inward angle. In fact, you just draw a line from the center of the rear wheel straight through the center of each head tube, as shown in Figure 7-9, and you get a fairly good angle to set the plates; no rocket science degree needed. If you really want to get into depth on the Ackerman steering concept, then search for "Ackerman steering" on the Internet and you will be able to read pages of information on the subject. If you just want to get

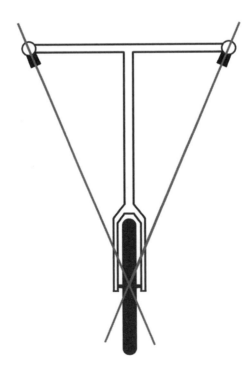

Figure 7-9   Find the angle of the two small plates by drawing a line to the back wheel.

up and running, trust me; this system works well and produces a steering system with almost no tire scrub or odd handling side effects.

Now, before we can weld these plates on, we have to make them, so get your hacksaw or cutting torch ready.

MAKING THE PLATES

Cut and drill the following parts from your 1/8 inch or similar thickness steel plate or flat bar using the measurements in Figure 7-10. The rounded area ground out of the top of each small plate is there to make a better joint when it is welded to the forks and should be made to fit. If your forks are flat in between the two legs, then you do not need to make this rounded area.

The holes in each plate must allow the bolts on the ball joints to fit through, and they should be snug, not oversized, or you will end up with sloppy steering. Also, make sure the ball joints are not badly worn and loose. Bad ball joints will make your steering erratic, or the system may come apart, causing your wheel to lock sideways. The holes are drilled in the plates as follows—on the large plate, the two holes are ½ inch from each corner; and on the two small plates, the holes are ½ inch from the bottom, and centered left and right.

Once you have the three plates cut out, weld the larger plate (the one with two holes in it) to the bottom of the stub on the delimbed center fork, as shown in Figure 7-11. The plate is welded so that the top (the end without holes) is flush against the top of the stub. All welding is done from the top, and the plate should end up at 90 degrees to the fork tube. Insert bearings and cups into the center head tube and put in the steering fork as it will be in the finished project. Also, connect one ball joint to the end of each rod and bolt them to the plate. If your ball joints do not thread into the rod, or you choose to use tubes like I

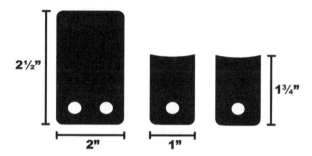

Figure 7-10 Cutting the plates that make up the steering system.

Figure 7-11   Center plate welded to the steering fork.

did, you will have to weld the ball joints to the end of the tubes or rods. Do not weld ball joints to both ends of the rod or tube yet, as you do not know how long they should be. They should be much too long right now, since you have pieces at least 12 inches long.

GETTING PROPER ALIGNMENT

OK, you may want to take a break or go inside and gear up your brain, because the next part is the last critical step, and must be done with accuracy in order to get your front wheels to go in a straight line.

Place both pairs of forks in the right and left head tubes, complete with bearings and all required hardware. Refer back to Figure 7-9 and look at how we find the proper angles to weld the two small plates to the right and left forks. As you can see, we need to draw an imaginary line from the center of each head tube back to the center of the rear wheel (between the rear dropouts). Turn the frame upside down and use whatever method you feel works best to position both pairs of forks perfectly straight, as if the bike were steering straight ahead. There are two methods I found that work well.

The first way is to tack weld a perfectly straight pipe or bar right across the front of both sets of forks, so that all four legs are tacked in perfect alignment. The other way is to drop a long, threaded rod across the forks dropouts where the front axle would connect; this will also put them in perfect forward alignment.

Once you are certain both pairs of forks are facing forward, align the center steering plate so it is also in the forward position. Now draw your line from the head tube centers to the back wheel hub (refer back to Figure 7-9), using a string or tape to get the two smaller plates' proper angles on the forks, and weld them in place, as shown in Figure 7-12.

Bolt the two remaining ball joints onto each plate on the right and left forks as they will be in the final design. We can now find the proper length of each tie rod by lining it up on the newly inserted ball joints and marking a line where they overlap. If you have the proper threaded rod and matching ball joints, then you don't have to be as critical with the cut, as you will be able to adjust your steering alignment by turning the ball joint bolt to set the distance.

Figure 7-12 Steering rods cut and welded at the correct length.

Of course, I did it the hard way, as you can see in Figure 7-12, and welded the ball joints to each end of the rod. The benefit to welding the joints to each end of the rod is that you can use any size ball joint and rod, and, if it is done properly, it will never need fine-tuning. The disadvantage, of course, is that the length of the rods is very critical. The benefit to using adjustable ball joints is that you can make a large error in cutting the rods and simply adjust the distance to compensate. The disadvantage is that it may work loose over time and need adjustment.

Once the rods are cut to the proper angle, make a small tack weld to secure the ball joints to the ends of the rod, making sure your forks are still in perfect alignment and all ball joints are bolted securely to the plates. When everything looks good, carefully remove the ball joints from the plates so you can finalize and grind the welds.

ADDING HANDLEBARS

Once you have all the steering components welded and installed, connect a pair of handlebars to the bike, and check to make sure it works. Notice in Figure 7-13 that the pair of forks on the inside of the turn are on a much sharper angle that the outer set. This is the magic of Ackerman steering.

If your steering system allows you to "oversteer," causing one set of forks to lock up or spin around, you may need to add a lockout system to stop this. I designed the steering so that the tie rod would hit against the middle fork stub if oversteered, but this depends upon the thickness of the rod, and you may need to weld a bolt or piece of steel in there to make this happen. Forty-five degrees of turn to the outside fork is plenty of steer to allow you to make circles on a narrow street. As for the handlebars, I chose a nice hefty BMX freestyle set with plenty of height, because I did not want to be hunched over on this bike as if I was on a racing bike. But the style of handlebars you use is totally up to you, or whatever parts you have on hand.

FINISHING TOUCHES

At this stage, it's a good idea to put on the front wheels and test the steering system. If your bike steers properly, then you are only a day away from taking over the streets once again! Take it all apart and get ready to paint your masterpiece, as shown in Figure 7-14. If you're worried about being visible in winter traffic, use a funky bright florescent paint, the type used for road signs or in construction work. Black

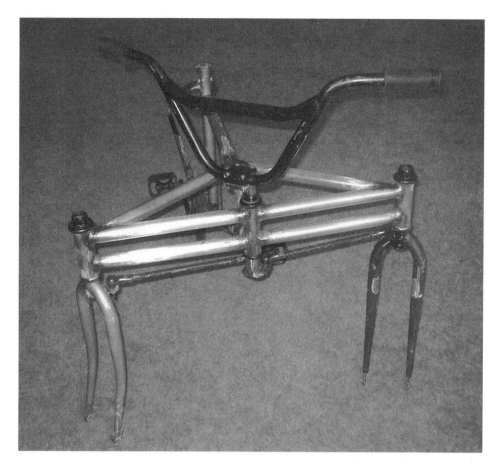

Figure 7-13 The finished frame with steering system installed.

would also be a good choice, as it will be in contrast with your white winter environment. Avoid the urge to bypass the painting stage and hit the trails, because exposed welded areas will decay and rust due to road salt and excessive moisture. You want your bike to look good, right?

Once your paint has dried and your machine is back together, you need to add the trimmings such as shifters, brakes, tires, and all that good stuff. I chose a massive knobby tire for the rear to get maximum traction, and medium-size front tires to get a good grip and help cut through the snow (see Figure 7-15). As for brakes, I only put them on the rear, for two reasons.

First, it would be hard to run a cable from the two front brakes to only one lever and, second, I wanted to be able to blast around a cor-

Figure 7-14 A nice coat of paint hides all those chunky welds.

Figure 7-15 Aggressive tires and good brakes help your machine perform well in winter conditions.

ner, lock up the rear wheel, and pull amazing t-slides and spin outs, and this only works with a rear-wheel brake. With only a rear brake, there is a decent amount of stopping power; but remember, it's winter, and you can't expect to stop on a dime, anyway.

My bike has 15 speeds and uses all of the original components from the main donor bike. I may also add a windshield made of Plexiglas® to the crossbar of the BMX handlebars and a bright light for night riding.

Riding the Hammerhead is not a lot different from riding a normal bike, except for the fact that you lean on the bike, not with it. If you are moving really fast, then crank the handlebars in one direction, you will end up on two wheels (a fun trick) unless you lean into the turn. Snowmobiles and quad runners have this same characteristic, and you get used to it in no time. In Figure 7-16, Devon Graham demonstrates the "wall of snow maneuver" by locking up the rear wheel while turning into a corner at high speed.

The Hammerhead will traverse just about any terrain, including snow banks, slush, steep hills (see Figure 7-17), icy roads, and hard-packed trails. Devon and I were riding over the frozen ice on the river

Figure 7-16 Learning to ride the Hammerhead is easy and fun!

Figure 7-17   Brad blasts over a snow bank at the top of a riverbank.

the day the picture was taken, and practicing sliding maneuvers on a long patch of ice. Driving the Hammerhead down steep hills made for sleds is also great fun, but watch out at the bottom of the hill, as you can still fly over the handlebars if you hit a large enough pile of snow, and this may or may not be a good thing, depending on your riding style!

The Snow Bus

Once the Hammerhead project was completed, I realized that I had created a very functional and fun winter vehicle. We would go to outdoor hockey rinks or icy parking lots and take turns spinning out and doing all sorts of fancy maneuvers on this almost unflappable machine. We even tied a sled to the seat post and pulled each other around. This worked surprisingly well wherever there was a decent layer of snow.

Although I could find no fault at all in the trike's design and handling characteristics, I thought that there was one more thing I wanted—another trike! Now I could ride all winter long on any surface and in any weather, but I had to do so by myself, since none of my usual riding partners had such a hardy machine, and a two-wheeler was no match for the Hammerhead.

Instead of building a complete trike all over again, I decided that it would be more fun to add an extra set of pedals and seat to the existing machine, turning it into a totally unique vehicle—a tandem trike with two front wheels, something you don't see every day, or any day for that matter. Adding a second bike to the original trike would be easier than making a whole new one, because the front end was the most complicated part to do, and the expansion would require only a minimum of easy-to-find parts.

SNOW BUS CHARACTERISTICS

I decided to call the new machine the Snow Bus, since "Hammerhead Tandem" did not sound very cool, and we all know, a cool name is a must for any project. Also, the bike is long like a bus, and when I take it out for solo rides, I can't get much farther than a block or two before someone asks for a ride and jumps on the back. The Snow Bus is an extension of the Hammerhead, so, obviously you have to make the Hammerhead before continuing with this project.

CUTTING THE FRAME

Now that you have built the Hammerhead, chop off the rear end with a hacksaw, as shown in Figure 7-18. After doing a lot of work, now you have to chop it in half! Even if you were planning to make the Snow Bus right from the beginning, it would still be best to start off with a single-passenger trike at first, then cut it up after it was working well. Following this process makes it easier to get things working correctly, especially with such a complex front end.

Add the removed rear end to your growing parts pile for later use, and dig out a donor frame to be used as the new rear end and passenger addition. You will be using only the parts on the new frame from the seat tube to the rear dropouts, so it doesn't matter if the forks or main tube are bent; they will be removed anyway. The donor frame

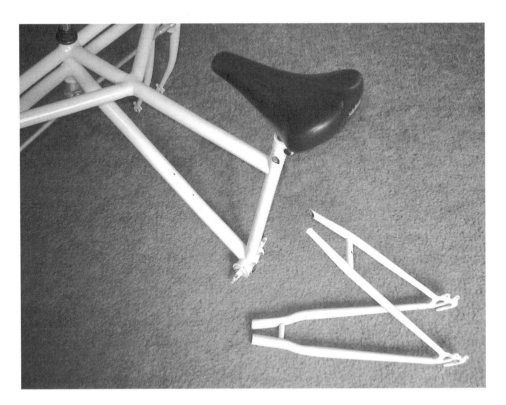

Figure 7-18   Cut the rear end off the original Hammerhead frame.

does need to meet a few requirements, though, such as wheel size and general geometry.

If you used a 26″ bike and forks for the front of your Hammerhead, make sure your addition also has the same-sized parts, or you will have an odd machine (not that there's anything normal about it already). Also, the frame height (length of seat tube) needs to be similar to the original or slightly smaller so that the new top tube will fit between the two seat tubes on each frame. For my donor frame, I chose a ladies' road bike that was the same size and height of the one I used to make the Hammerhead. The advantage of a frame like this is that the lower main tube allows easy access onto the bike for the stoker (the rider who sits at the rear), and follows the general line of the original main tube for a nice-looking frame.

The disadvantage of this type of frame is strength. If the stoker is 6′5″ and 230 pounds, you may want to use a heavier men's style frame. Also, the extra set of stays in between the seat stays and chain stays on my chosen frame are very necessary for strength because a lot of weight will be transferred along both main tubes and into the joint at the rear seat tube.

Once you have chosen a suitable frame, cut the front end off as close to the seat tube as possible (see Figure 7-19) and add it to the scrap pile. Take note of the original length of the main tube, as you will need to keep this measurement the same for the new main tube and boom (lower tube that joins the two frames) so that the stoker is placed at a reasonable distance on the bike to allow for a comfortable ride.

If the stoker's seat is too close to the captain's seat (front seat), then there will be a constant collision between the stoker's knees and the captain's buttocks (not fun). If the distance is too far, the stoker will have to hunch over to reach the handlebars, causing a collision between the stoker's face and the captain's back (also not fun). As shown in Figure 7-19, the extra tube is a piece of one-inch electrical conduit that I used to joint the frames, but the original tubing from the donor bike would also be good if this is all you have.

MAKING THE BOOM

It's time to cut the tube that will be used as the boom to the correct length. Use the measurement that you took from the donor frame's main tube, or measure the main tube on the front frame. Because both seat tubes should be parallel in the finished trike, the boom and main tube must be the same, or very close. The tube I cut was 17″. Before you weld the boom, it's a good idea to set up the two frame halves so

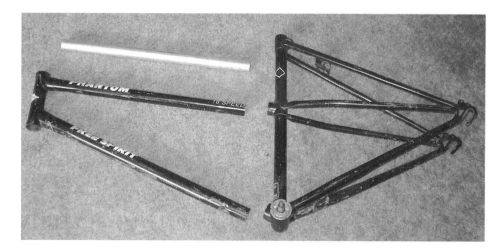

Figure 7-19 Cut the front end from the donor frame.

you can get a good idea of how the final product should look, as shown in Figure 7-20. Notice the plastic container under the rear dropouts. The front forks are resting on the ground, so the rear dropouts should be elevated, as they would be if a regular bike were placed on the ground with the front forks down.

There is about a 4″ distance between the rear dropouts and the ground if a frame is resting on the fork tips and bottom bracket. The same distance is needed for this tandem. If you welded the two frames together without this 4″ difference, it would still work properly, but the two seat tubes would be at odd angles, and the captain's frame would be leaning forward. The goal here is to have both seat tubes at the same angle.

Once you have the two frames laid out properly with the new boom resting between them, tack weld the boom at the top on each side then carefully align each frame as straight as possible, taking a good look at every angle—front, back, and sides—and making slight adjustments without breaking the weld. Both frames need to be as straight as possible in order to end up with proper tracking and chain alignment, so do this carefully.

Figure 7-20   Cutting the tube to be used as the boom.

Weld only a little on each side to help prevent warping as you weld the boom in place. Make any adjustments needed after each small weld has cooled, and continue this process until the boom is welded all the way around. Once the boom has been properly welded, clean and grind the welds before moving on.

MAKING THE REAR MAIN TUBE

The next tube to be cut and welded will be the new rear main tube (see Figure 7-21). This tube should be almost the same length as the boom, since both seat tubes are parallel to each other. Take this measurement and add an inch or so for the rounded groove that will be cut at each end, to allow for a good-fitting joint, and cut the tube. I used conduit again for this tube, but any bike tubing will work fine here as well. Again, make sure both frames are straight before welding-in the tube. You can still correct the vertical alignment with a good swat from

Figure 7-21   Measuring the length for the new rear main tube.

a rubber mallet at this stage, but once that new main tube is welded in place, you have to live with any errors, or take out the hacksaw again, so take your time and get it right.

Weld in the rear main tube using the same method as for the boom, taking care to control the heat and avoid warping. Fill all holes and imperfections and grind the welds. There's only one more tube to weld before you have a completed frame.

Now that both halves of the frame are welded together, it's a good idea to make sure that there are no serious alignment problems. Put all three wheels back on the trike, as shown in Figure 7-22, and have a good look at it from all angles, especially from the rear. The two seat tubes should look parallel from the side, and should be in perfect align-ment from the back. Also, the two bottom brackets should be parallel when viewed from the top.

If one of the bottom brackets is severely out of alignment, the chain may not stay on due to alignment problems. Do not be tempted to sit on the bike yet, as the frame is not ready to support any real weight. The problem is that the square formed from the two seat tubes, rear

Figure 7-22 Checking frame alignment with the wheels attached.

main tube, and boom is not structurally strong, and needs to be altered before it can hold any weight.

MAKING THE TRUSS TUBE

Imagine compressing an empty picture frame by pressing on two diagonal corners. It would, of course, fold with very little effort. If we added another piece between those two corners, it would not bend at all, since you would have to crush the new piece of the frame as well. To make our trike frame strong, we will add this extra tube (truss) between the two corners of our square. The truss tube will run between the joint at the front seat tube and main tube to the joint at the rear seat tube and boom (see Figure 7-23). This tube can be lighter than the

Figure 7-23 A truss tube is added to give the frame strength.

main tube (I used 3/4-inch conduit), but any tube that you have will also work fine. Remember to add a little extra to the measured distance to compensate for groove cuts before welding.

ADDING CRANKSETS

Now that your frame is complete, you only need to add the components and you will soon be able to try out your new machine. Begin by adding the two cranksets to the trike. Although both cranksets do not have to be identical, there are certain requirements that have to be met for everything to work properly.

First, the smallest chain ring on each crankset needs to have the same number of teeth. Since these are the chain rings that will join the captain's crankset to the stoker's crankset, they must have the same number of teeth in order to keep the pedals in sync. If one of the chain rings had even one extra tooth, eventually the pedals would fall out of sync, and a collision of feet would occur.

Second, at least one crankset must have two chain rings. The rear set needs to have two chain rings since it will deliver power to the rear wheel. The dual chain ring at the back is necessary because there are two separate chains. The front chain transfers power from the captain's crankset to the stoker's crankset and keeps both cranks synchronized. The rear chain then transfers all the power from both cranksets to the back wheel.

Using a single long chain poses a problem. If a single chain were used, only a few teeth would mesh with the top and bottom of the stoker's chain ring as the chain passed by. The chain would jump right off the chain ring if any pressure were applied to the stoker's pedals because a chain needs to cover at least half of a sprocket in order to function properly. I did try using a single long chain once, but you shouldn't bother—it doesn't work!

Now that you understand what we are trying to achieve with this chain, make a chain long enough to ride on both smaller chain rings. This chain shouldn't be so tight that it can't be put on by hand without force. The chain should also not be any longer than the minimum length that will fit. In the end, you will have a chain that has a little slack, as it will be slightly too long (see Figure 7-24).

THE DERAILLEUR

Once your connecting chain is in place, you need a way to pick up the slack to stop the chain from falling off. If you spin one of the cranks

Figure 7-24 The front chain is added between the two small, equal-sized sprockets.

right now, it won't take long for the chain to bounce off one of the chain rings, and this would not make for fun riding. Chain slack is not a new problem, nor is it a hard problem to fix. Look at any bike with more than one speed and you will see a device designed to deal with that problem—the rear derailleur. The two small plastic wheels and spring mechanism on the rear derailleur eliminate chain slack by pulling the return chain in to keep it tight at all times.

There is no reason to reinvent the wheel, so we will just use an existing rear derailleur to pick up the chain slack. The derailleur is welded to the boom so that the two plastic wheels are placed in line with the chain and in the center of the boom (see Figure 7-25). Notice how the front side of the derailleur (the side that you normally see when looking at the gear side of a bike) is facing the opposite way in our configuration.

To get the chain into the derailleur, remove the bolt that holds one of the plastic wheels on, and put the chain in the middle of the two. Once

Figure 7-25 An old rear derailleur picks up the chain slack.

the derailleur is welded to the frame, you can turn one of the two small adjusting screws in order to further tweak its alignment between the two chain rings.

After welding the derailleur in place, you should be able to spin either crankset wildly without the chain hopping off either chain ring. If there is any rubbing between the chain and the metal frame of the derailleur, try adjusting the position screws. If this fails, you may need to remove the derailleur and try welding it in a different position. When everything is set up properly, there will be very little friction and noise, and no rubbing of the chain against any part of the derailleur.

The rear chain is no mystery. In fact, it is no different than on an ordinary bike. The chain will be placed on the larger rear chain ring, then through a standard rear derailleur to the back wheel, as shown in Figure 7-26. If your rear crankset has the chain rings really close together, the connecting chain and rear chain my rub together. This can be solved in one of two ways. First, most cranksets have three chain rings—a really small one (the one we chose to connect stoker to cap-

Figure 7-26   Rear chain connected in the usual manner.

tain) and two larger ones. If your crankset has three rings, use the smallest and largest, leaving one empty in the middle. This will allow plenty of distance between both chains. The second option is to take apart the chain rings by removing the bolts or rivets, and add a washer in between both chain rings to keep both chains from rubbing. The first method is much easier.

The only other concern with the rear derailleur is cable length. Now that your trike is 2 feet longer than a normal bike, a standard shifter cable will be too short. You can either go to your local bike shop and have one made, or mount the shifter at the rear so the stoker can operate it. The first solution is the better one since you may want to ride solo, and reaching back to shift is not cool.

ADDING THE REAR HANDLEBAR

You almost have a fully functional two-headed, all-weather, tandem mountain trike now, and all you need to do is throw on the rear seat and go for a test ride, but what is your poor stoker going to hold onto? Maybe around your waist? Well, this may be good if you're on a date, but not too good if you are with your buddy, so we need to make some type of rear handlebar. You could cut a pair of handlebars in half and weld them right to the front seat post, but this would not allow for any adjustment. A simple but effective solution is to cut the end of an old gooseneck, as shown in Figure 7-27, then weld it to the front seat post. This will allow you to use different handlebars, as well as adjust their position front and back.

When choosing a pair of stoker handlebars, choose a wide and tall set, similar to ape hangers, or a BMX-style set (with the crossbar cut out), as the captain will have to sit in between them. If your stoker bars are too low, the stoker will have to bend over to reach them, or they may interfere with the knees while pedaling. If they are too narrow, the captain will not be able to sit on the seat because the bars will be in the way. The easiest way to find a good set is to weld the neck piece to the seat post, sit on the bike, and try them out until you find the perfect set.

The end of the gooseneck is cut from the stem with about an inch to spare. It is then welded just under the junction between the thin and thick part of the tube in order to make room for the seat. In order to

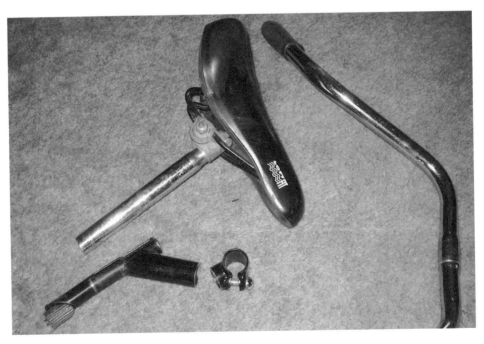

Figure 7-27   Parts needed to mount the stoker's handlebars.

make sure the handlebars will be straight, tack weld the top at first, so you can make adjustments to the angle, then tack the bottom after everything is straight. Weld all around the joint, making sure the weld is strong and free from cracks (you don't want to lose your stoker), then grind the weld.

When the seat post is mounted to the bike (see Figure 7-28), the stoker should be in a good position and be able to pedal without hitting the bars or captain's seat. The captain should also be able to sit on the seat without having to squeeze in between the handlebars.

THE TEST RIDE

Now for the part you have been waiting for—the test ride (sometimes called the "crash test"). Tighten all nuts, align both cranksets so the pedals are synchronized (see Figure 7-29), and get out your helmet.

First, take the trike out by yourself and get a feel for the handling. You will notice that the trike actually feels more stable and may have better traction through deeper snow. This is due to the longer back end dividing the weight more evenly across all three wheels. Other than

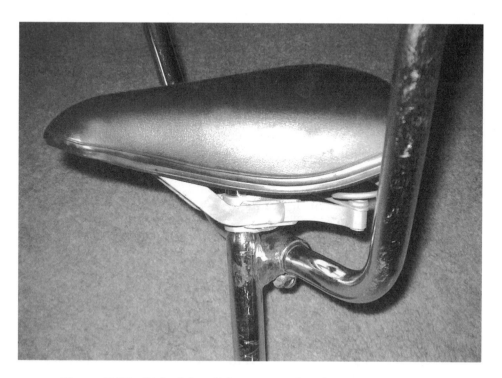

Figure 7-28   Stoker's handlebars mounted to the captain's seat post.

Figure 7-29   Completed Snow Bus ready for a test ride.

that, the machine should feel like it did before you cut it in half to add the second frame. Once you know the bike is in working order, find a stoker and see how it rides.

Make sure you tell your passenger the rules—lean when I lean, stop pedaling when I stop pedaling, and jump when I yell "Bail!" Leaning together into fast corners is important, and you will soon see that what the stoker does can have just as much effect on the bike as what you do. Teamwork is important.

FINISHING TOUCHES AND TIPS

How did it handle? Did the chain stay on? Did the frame snap in half? Chances are, everything went well and both captain and stoker had a great time. Now, you need to rip it all apart again, and paint the back three-quarters of the bike (see Figure 7-30). Don't be tempted to skip the painting stage, as this vehicle will be subjected to harsh elements such as slush, road salt, and sand.

Figure 7-30 Use a good paint for the Snow Bus to protect it from corrosion, road salt and moisture.

Use a good paint, and give it a little extra on the underside since this will be an area getting constant abuse from moisture. Also, you may want to leave the trike outside if you plan to ride often, since the warming up process will cause a lot of condensation, leading to increased rust.

Once captain and stoker are working together as a team as shown in Figure 7-31, the Snow Bus is a fast, fun, sociable vehicle, great for beating those winter blahs. It's surprising how fast you warm up even on those unbearable –30° nights that drive even the heardiest winter lover running indoors. With a full-face mask, two layers of clothes, and warm gloves, you can ride for hours and stay at a very comfortable temperature due to the aerobic effect of pedaling. The vehicle is also surprisingly fast, even on snow or rough ground. When both captain and stoker are pedaling hard and in sync, use defensive driving meth-

Figure 7-31 Captain and Stoker must work together, especially going around corners.

Figure 7-32 Brad and his dad, Tom, hitting the trails on the Snow Bus.

ods and be wary of reduced stopping power on icy surfaces. Think ahead as much as possible and drive wisely.

I am glad that I started this project, as it really turned out to be a fun and useful vehicle. I use the Snow Bus for both exercise and entertainment, and it has taken a lot of punishment without any problem. For keeping in shape, a 30-minute or one-hour ride is great (see Figure 7-32), especially since plowing through the snow is more work than riding on a flat surface in the summer. If you really want to build up your riding stamina, tie a rope to the seat post, and pull someone on a sled for a few blocks if you can. This is a lot of fun and a real leg blaster as well! We have conquered everything from hockey rinks to steep hills and always find something new to do on the trike each time we go out. And we always have a blast.

THE SKY IS THE LIMIT

The Inverta-bike

Take your cycling experience to new heights with a tall bike! Although the Inverta-bike is not nearly as tall as the Skycycle project that is outlined later in this chapter, it can still dwarf any normal upright bicycle and put you up high enough to look a trucker square in the eyes. This project uses a minimum of parts and can be built in one evening from just about any scrap tubing you may happen to have in your junk pile.

Riding a tall bike like the Inverta-bike is an entirely different cycling experience, and you almost feel like you are flying because you don't see much of the bike beneath you as you ride, only the ground passing by. If you don't like to draw attention to yourself while riding, then this is not the project for you, as you will turn heads and drop jaws in confusion and amazement wherever you ride. Although a tall bike may look dangerous and hard to ride at first, once you are in the driver's seat you will realize that it is no different than an ordinary ground-hugging bicycle, and it may even offer added safety features such as the ability to see traffic for miles ahead and avoid ankle-chomping dog attacks!

FINDING THE RIGHT PARTS

Creating the Inverta-bike is fairly simple and requires a working bicycle of any shape and size and a few feet of one-inch steel tubing. Your donor bike does not have to be in great shape. It can be of any decent quality, as long as it can still be ridden and has equal-sized wheels front and back. For my design, I chose a heavy, old mountain bike scavenged from the scrap yard (see Figure 8-1). Any frame size will

Figure 8-1 Donor bike for the Inverta-bike should at least be in working condition.

work, from 28 inch right down to those little kid's bikes with the 12-inch wheels. However, the larger the donor frame, the taller the final product will be.

Once you have found a working bike to use for this project, remove almost everything because we will be putting all the parts on backward. After you have removed everything, including both crank arms, you must remove one of the end caps on the bottom bracket and turn the crank axle around the other way, as shown in Figure 8-2. The reason for this is to allow the longer side of the axle (normally on the right of the frame) to be placed on the left side of the frame to accommodate the reversed chain ring side crank arm. Yes, one side of the axles is longer than the other to allow extra clearance between the chain ring and chain stays. If you did not reverse the axle, the chain ring would

Figure 8-2 The crank axle must be reversed so that the long side is on the left side of the frame.

rub on the frame and act like a hacksaw, and you will have invented a "self destructing" bike that would saw itself in half.

The forks will also be turned upside down, so you will have to bang out the fork bearing cups and reverse them as well, since one is usually larger than the other. The larger bearing cup should be at the bottom of the bike where the forks enter the frame, but this is now the top of our bike since we will be turning the frame upside down. Confused yet?

Take a look at Figure 8-3 and it should make a little more sense. As you can see, just about everything is on backwards—the frame is upside down, the forks are on the other side, and the crank arms are reversed. The only part of the bike that could not be reversed is the rear dropouts (actually, they are now dropins).

THE REAR DROPOUTS

Because the rear dropouts are pointing the wrong way, you will have to make sure that the rear axle nuts are on good and tight or you may not be going very far!

Although you could actually ride this contraption as it is right now by standing on the pedals, hunched over the low handlebars, it will probably be a better idea to make a seat tube.

THE SEAT TUBE

A suitable seat tube can be made from any tubing or even an actual seat tube cut from another bike. In my design, I used a piece of one-

Figure 8-3   Frame and forks are now turned upside down, and cranks are reversed.

inch electrical conduit, because I have miles of it laying around at all times, and a regular seat post fits nicely inside it. Cut a piece of tubing to be used as the seat tube so it is a few inches shorter than your inseam (length of the inside of your leg). This way, the seat post can be adjusted to suit your height. You will also need a seat post clamp and seat post (see Figure 8-4).

To allow the seat post to adjust, a slot is ground from the seat tube with a grinding disc or hacksaw so it resembles the one shown in Figure 8-5. The clamp is then placed back over the slot and tightened. Of course, if you just used a seat tube cut from another frame instead, you can skip this step. If you plan on making this bike for multiple riders, make sure the seat can go as low as it needs to for the shortest rider.

When you have finished making your seat tube, it will be welded on top (formerly the underside) of the bottom bracket (see Figure 8-6). When you are welding this tube in place, put both wheels on the bike, inflate the tires, and sit the bike on level ground so you can position the tube at 90 degrees to the ground. Not only should the seat tube be perfectly vertical, it should also be in alignment with the rest of the frame, so only tack weld it at first. Look down the length of the bike from the front or rear so you can get the seat tube in perfect alignment before you add the final welds at the bottom bracket.

The seat tube will not be very strong by itself because it is only welded at the base where it meets the bottom bracket, so some type of gusset or support should be added. I used a premade, one-inch electrical conduit elbow, as shown in Figure 8-7, but any piece of tubing, straight or bent, will do the job. Feel free to get fancy here and use your imagi-

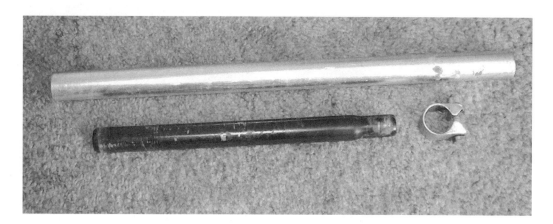

Figure 8-4 Cut the seat tube a few inches shorter than your inseam.

Figure 8-5 Seat post and clamp placed into the seat tube.

Figure 8-6 The seat tube will be welded onto the bottom bracket.

Figure 8-7   The seat tube will need a support tube or gusset to make it stronger.

nation. This style tall bike has been made in hundreds of different configurations throughout the years.

Try to avoid using any pieces of tubing or parts that stick up or have sharp edges. If you have to get off this bike in a hurry, or against your will, you don't want to land on a sharp piece of tubing or gusset.

THE GOOSENECK

The last thing that has to be made is the long gooseneck, unless, of course, you have five-foot-long arms to reach all the way down to the forks! The easiest way to extend the gooseneck is to cut it in the middle and add a length of tubing in between. Choose a gooseneck made of steel, not aluminum, so we can weld to it. Also, choose a one-inch tube that is fairly strong, so that there is minimal flex in the final product (one-inch electrical conduit works well). To figure out what length to make the extension tube, throw on the seat and adjust it to the correct height, and then measure the rough distance between the fork top and the position you want the handlebars to be at in the final design.

Cut the gooseneck in the middle so that the cut is parallel to the stem (see Figure 8-8). The gooseneck must be made long enough to place the handlebars at the proper height and position for the rider. Try to keep the added tube at roughly the same angle as the original gooseneck's main stem, or you will find that the handlebars will swing from side to side when you are turning. This is called a "tiller effect."

When you have cut the neck in half, grind out the ends of the cuts so that they create a tight fit with the extension tube, this will make a stronger, better-looking final weld.

The best place to start is by welding the top of the gooseneck to the extension tube, as shown in Figure 8-9. This part is less critical because the vertical angle will not really affect the position of the handlebars in relation to your body. But do make sure that the horizontal angle is as close to 90 degrees as possible so your handlebars are not misaligned.

To get the horizontal angle as close to 90 degrees as possible, mount a set of straight handlebars into the gooseneck once it is tack welded to the top of the extension tube, then you can make it straight using the handlebars as leverage to force it straight as you weld. Make a good job of these welds so your extended gooseneck does not come apart while you are riding.

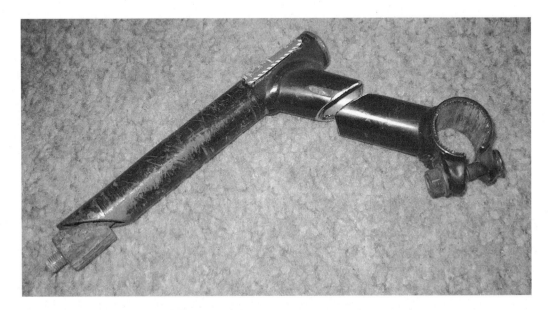

Figure 8-8   Steel gooseneck is cut in the center, parallel to the stem.

Figure 8-9 Top half of the gooseneck is welded to the top of the extension tube.

The extension tube will now be welded to the bottom part of the gooseneck. You will want your handlebars to be a reasonable distance from your body, as they are on a normal bike, so place the bottom half into the fork stem before you start welding. This way, you can just tack weld the extension tube in place and make adjustments as you finalize the welding. Leave a little extra tubing beyond the bottom of the cut in the gooseneck (see Figure 8-10). This will stop any burnthrough from happening on the end of the tube. Again, try to make this weld as strong as possible, as it is your only interface to the bike besides the seat, and I can't imagine that it would be fun to ride this bike without any steering!

FINISHING TOUCHES AND RIDING TIPS

After the extended gooseneck has been assembled, all you need to do is put on a chain, a seat, and a helmet for your first test ride. I removed the rear derailleur and set the chain on the third rear chain ring so it would be a fixed-gear bike of moderate speed. Just use your chain link tool to make the chain the correct length and set the rear wheel in

Figure 8-10 The bottom of the gooseneck will be welded to the extension tube.

place so that the chain has no slop. Notice that the gooseneck extension on the finished bike, as shown in Figure 8-11, is at almost the same angle as the actual gooseneck, putting the handlebars in a position similar to that of a normal bike. If you are going to use a cable brake in your design, put it on the rear wheel, not the front. Do I have to explain why? I didn't think so!

Figure 8-11 Final product ready for a test ride.

Riding the Inverta-bike is fairly simple. Hold the handlebars, then put your outside leg on the pedal while you coast the bike up to speed. Grab hold of the seat with the other hand and swing your inside leg over the seat. Have you ever seen a cowboy jump onto a horse? Getting on this bike is similar. Once you are riding, everything is the same as a regular bike. Your turning circle may be a little wider due to the fact that you can't lean as far into the corners, so give yourself ample room if doing a tight corner or turning a circle in the middle of the street. Figure 8-12 shows Devon Graham having a good time cruising down the street on the newly painted Inverta-bike.

This simple tall bike design can be easily modified to be taller just by adding another frame on top of the existing frame, creating a "double decker" style frame. In fact, I have seen bicycle hackers even build tall bike with three or more frames stacked on top of each other!

If you get bored with only being six feet in the air and crave now heights, then get ready for the Skycycle project presented next in this

Figure 8-12   Devon Graham sails down the street on the Inverta-bike.

chapter. The Skycycle can be built to just about any height you are brave enough to climb, and might be just the thing to satisfy your need to fly.

The Skycycle

Standing over 10 feet tall from the ground to the seat, the Skycycle is truly the king of the tall bikes. Riding this high-altitude vehicle is more like piloting an aircraft than riding a bike, as you see only the handlebars and the road ahead of you from such a high vantage point. This bike is not a good project for those with a fear of heights and will even raise a few hairs on the backs of daredevils who are at home in high places. The frame of the Skycycle looks more like a construction ladder than that of a bike, and for good reason—so that you can climb up to the top! Once at the top, you just ride it like an ordinary bicycle (if you are brave enough). Riding the Skycycle is not any more difficult than riding a regular bike, besides the fact that you have to duck under low tree branches and avoid places where you have to stop completely. To get off the Skycycle, you just grab hold of a pole or wall, and then climb back down the ladder-like frame. Of course, there is another way down (jumping off), but I try to avoid that method unless an emergency situation comes up. The Skycycle can be built to just about any height you like but, personally, I think 10 feet would be as tall as you would ever need, and I doubt you would be brave enough to make a taller unit. (Is this a dare?)

THE BASIC FRAME

As shown in Figure 8-13, the basic frame for the Skycycle is made like a ladder, with the seat tube and head tube forming the rails. Although you can make this frame as tall as you like, there are some general rules that should be followed, such as the angle at the joint of tubes "A" and "B" should be exactly 90 degrees, and tube "B" should be centered between the front and rear wheels. By following these rules, your Skycycle will be properly balanced and ride much like a regular "down-to-earth" bicycle.

Before you head out to the garage and start chopping tubes, take a look at the rest of this chapter to get an idea what you will need to do to finish this project.

Most of the frame is made from 1.5-inch, thin-walled electrical conduit, with the exception of the long fork stem that is made from one-

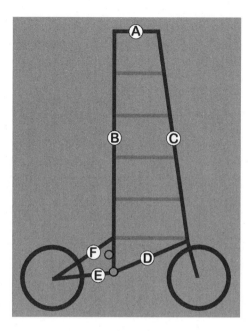

Figure 8-13   Basic frame geometry of the Skycycle project.

inch electrical conduit. You will also need three bottom brackets salvaged from some old bike frames, three sets of cranks to fit the bottom brackets, a set of forks, two wheels, a seat, and a whole lot of chain. For the rear wheel, a coaster brake hub is used.

For the very long chain, you can join many bicycle chains of equal size together, or use a few garage door opener chains if you can find them, as they are the same size as BMX chain. When you are selecting chains for use on this project, make sure they are all the same size and free of bends and rust. It is important to keep the chain from falling off this bike, or you will be jumping off rather than climbing down the easy way.

1.5-inch exhaust tubing will also work for the main frame tubing, but is a lot more expensive than electrical conduit, and made of basically the same material. Avoid any really heavy tubing like that used in plumbing or construction, or you will end up with one really heavy frame. With the conduit used for the frame, it will weigh about 70 pounds if made 10 feet tall.

CUTTING THE TUBES

The first thing you need to do is choose a length for tube "B," the kingpin tube of the frame. This tube will determine the overall height of your Skycycle and connect all the other tubes together. In Figure 8-13, you will notice that this tube starts at about the same height as the wheel hub and stops at the base of the seat post. If you made this tube 8 feet long, your total bike height would be about 10 feet tall after adding the height of half a wheel and a seat post to the length. Don't go overboard when choosing a length for tube "B" or you will create a bike too scary for even the bravest daredevil to ride.

If you think you will be able to ride a 10-foot-tall bike, then go find a ladder of equal height, climb to the top, and sit there. You may realize that 10 feet is higher than you thought, especially when not on solid ground. Five to 6 feet is a pretty good length for tube "B," 6 to 8 feet is getting brave, and anything longer is insanity, but it's your call.

Once you have chosen the length for tube "B" (remember your final bike will be two feet taller), cut it square at both ends. Now you will need to cut a length for tube "A". Tube "A" will become your top tube, and should be the same length as it would be on a regular bike. A length of 19 inches is about average and will work just fine. If you are a very tall rider, than add a few inches extra for leg clearance so your knees don't hit the handlebars.

Once both tubes ("A" and "B") have been cut, they are welded together (see Figure 8-14) so there is an exact 90 degrees between them. Leave a half an inch at the top of tube "B" as well, so you can weld the seat post plate on top later.

The next tube to be cut will be tube "D," the down tube. This tube will determine the length, or wheelbase, of you bike. If the wheelbase is made too short, the bike may tip forward or backward under extreme conditions such as climbing a steep slope or riding over a tall curb. If the wheelbase is made too long, you will not be able to turn around in the width of a city street. I have found that the following lengths worked out well for stability and turning radius:

- For a bike that will be 6 to 8 feet tall, use a length of 28 inches for tube "D."

- For a bike that will be 8 to 10 feet tall, use a length of 30 inches for tube "D."

- For a bike that will be 10 to 12 feet tall, use a length of 32 inches for tube "D."

If your bike will be over 12 feet tall, then you have boldly gone where no cyclist has ever gone before, and I cannot help you!

Figure 8-14 The joint between tubes "A" and "B" must be exactly 90 degrees.

Now that you have tubes "A," "B," and "D," it is time to lay out your frame on a flat, level surface so you can determine the placement of everything in relation to the wheels and also to get the proper length for tube "C."

The size of the wheels you choose will determine the angle of tube "D" and the length of tube "C," since they must be welded together above the front wheel. Remember, the goal of the final design is to have tube "B" reach the stars at exactly 90 degrees to the ground, so your weight is placed equally between the two wheels. Lay out your tubes as shown in Figure 8-15. Remember that tube "B" is always 90 degrees to the ground, and there is an equal distance between the wheels and tube "B."

To get the proper length for tube "C" and angle for tube "D," put the wheels in place so that the bottom bracket (placed at the end of tube "B") is at about the same height as the wheel axles. The front wheel

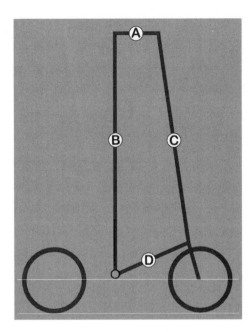

Figure 8-15   Laying out the frame tubes to get the proper angles.

should have the forks attached and in place in the position they will be when the frame is complete. Now, with tube "D" at the correct angle above the front wheel, you can measure the proper length for tube "C," the head tube.

Tube "C" will stick up past tube "A" about a half inch, just like tube "B" does (the seat tube). This extra half-inch makes it easier to weld the end without a burnthrough. The same applies at the bottom joint between tubes "D" and "C."

GROOVING THE TUBES

When you are satisfied with your layout, cut the tubes, leaving enough extra length to compensate for the grooves you will have to cut in order to make a tight joint (see Figure 8-16). Don't worry much about high precision when determining the length of all the tubes; just make sure the two rules are followed—the seat tube (tube "B") must be 90 degrees to the ground in the final design, and the front and rear wheels should be an equal distance from either side of this tube.

If you follow these rules, your bike will stand up perfectly straight, and you will be centered between the wheels as you ride.

Figure 8-16 Cut tubing to an extra one inch to allow for grooving the joints.

When you have grooved out all of the tubes to complete the main part of your frame, lay it all out again on a level surface, with the wheels in place just to make sure everything looks alright. The bottom joint between tube "B" and tube "D" will contain a bottom bracket, so cut out whatever amount of tubing you need to fit it in place.

Once you have all four tubes cut and the bottom bracket in place, your frame should look like the one shown in Figure 8-17. Notice the extra half-inch of tubing at end of the joints on the head tube (tube "C") to allow for placement of the fork bearings. Tack weld the frame as it lays on a level surface, then stand it up for observation. If there is any misalignment in the frame, twist it back into place as you add more welds to each tube. The final frame should be aligned both horizontally and vertically, so take your time and check it as you weld.

ADDING STEPS

Now that you have the basic frame perimeter complete, you need to add the steps in between tubes "B" and "C" so that you have some way of climbing up to the seat. Because the topmost step also carries the

Figure 8-17 The basic frame, including lower bottom bracket, ready to be welded.

main bottom bracket, you must choose a measurement for the spacing of the steps that fits the frame height and your leg inseam. If you put the first step down too far from the top of the frame (tube "A"), then you will never be able to reach the pedals. I chose a distance of 16 inches between the steps, as this would accommodate a large variance in rider heights just by adjusting the seat post, and it divided up nicely into five steps.

I recommend that you place a seat into the top of tube "B," the seat tube, and measure the distance from the top of the seat to the farthest pedal just to make sure the distance will not be too great for the shortest rider's inseam. Hold a crankset in place over the top step, as shown in Figure 8-18, as if it was welded in place. Placement of the crank on the top step set is not critical, but if you keep it in a similar position relative to the seat as in a regular bike, riding the Skycycle will feel a little bit more "normal," if that is possible! Five or six inches from tube "B" (the seat tube) seems to be about right.

Once you have determined a good spacing for the steps, just mark these points on tube "B," then take a 90 degree measurement from

Figure 8-18 The position of the top step also determines the minimum inseam distance.

each point on tube "B" to tube "C." When you are done, each step should be parallel to the top tube (tube "A") and to the ground. Don't forget to add an extra two inches to each length so that when you groove out each end, it will not end up too short to fit back into place. When you are done welding the steps, they should all be parallel to each other (see Figure 8-19). If you accidentally cut one of the steps too short, do not be tempted to "squeeze" tubes "B" and "C" together to shorten the gap, as this will distort your final frame. Instead, fill in the gaps with weld or cut a new tube.

THE LONG FORK STEM

Now you need to create the long fork stem that will slide into the head tube (tube "C"). First, the fork bearing cups must be placed at each end of tube "C" in order to get the proper length for the fork stem. The cups can either be welded into the ends of the tube, or you can cut an actual head tube in half, and weld the two pieces into each end of tube "C" to hold the cups. Both methods work well. If you choose to weld the cups directly to the ends of the tube, then only weld in short quarter-inch beads in order to minimize the warping of the cups from heat. Once the bearing cups are in place, measure the distance from one to the other along the length of tube "C," then add two inches to this length. The two extra inches are for the top hardware (fork nuts). The fork stem will be made from a one-inch length of electrical conduit. This works out well since the original fork stem fits nicely inside a piece of

Figure 8-19   The steps are welded in place so they are all parallel to the ground.

one-inch conduit. Before you cut any conduit, first take the set of forks you plan to use in your design, and cut them in half at the stem, as shown in Figure 8-20. The extension stem will be welded in between both halves of the fork to create the new ultralong fork required for the Skycycle.

THE EXTENSION TUBE

Look at how the bottom of the fork is joined to the extension tube in Figure 8-21. This is the reason why cutting the conduit was a good idea— you will have no way of telling how long the conduit needs to be until it is placed inside the head tube (tube "C"). Start with an entire 10-foot length of conduit, and place the bottom of the fork into the end (see Figure 8-21). Make sure the "bearing race" is in place on the fork before you do any welding, or you will not be able to get it back on since the conduit is much wider than the original fork stem. The bearing race is the small ring located at the base of the fork stem (see Figures 8-20 and 8-21). Make sure at least an inch of the old fork stem gets put into the tube as well, for added support and to help align the pieces.

Figure 8-20 Cut the stem of your forks in half so that they can be extended.

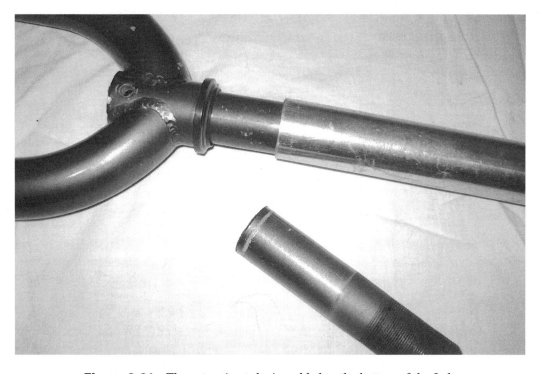

Figure 8-21 The extension tube is welded to the bottom of the fork.

Figure 8-22   The top of the fork stem is welded to the extension tube.

Once you have the bottom of the forks welded solid to the extension tube, place the unit alongside the head tube as if it were mounted in place. Now, you can decide where to cut off the top of the extension so when you weld the remaining fork stem (with the threads), the top of the threaded end will extend about two inches past the top cup so you can mount your hardware. Once you have found this magic length, mark the conduit, cut it, and weld in the top stem just like you did with the bottom of the forks (see Figure 8-22).

Once the forks have been completed, you can cut out the grooves in the frame for the two bottom brackets that need to be added—one on the top step to hold your crankset and the other just above the lower bottom bracket on tube "B" that will guide the top of the chain to the rear wheel. Because you've already figured out where the crankset bracket needs to be placed on the top step, this should be no problem; just cut out the groove like the one shown in Figure 8-23, and weld in the bracket, making sure it is aligned properly. Check your calculations one more time to make sure you can reach the pedals by placing a seat at the end of tube "B" and measuring from the top of the seat to the farthest pedal before you start cutting the groove.

The other bottom bracket that needs to be added will be placed just above the lower bottom bracket that is already in place so that when both chain rings are in place, they will not touch one another. This bottom bracket is also placed slightly to the left of the lower one so the drive and return chains do not rub against each other. Refer back to Figure 8-13 to get a better idea of the placement of the two bottom brackets. Once you understand how they are placed, cut the groove as shown in Figure 8-23, making sure that the alignment is as straight as possible.

Figure 8-23 The bottom bracket that holds the crankset will be welded to the top step.

INSTALLING THE TOP CRANKSET

Once the other two bottom brackets have been welded in place, install the top crankset. For the lower bottom brackets, two equal-sized cranksets with the crank arms cut off are placed into the lower bottom brackets. These will be used to guide the chain from the top crankset to the rear wheel. Without these two extra chain rings, the chain would run on an angle from the top of the bike to the rear wheel, and this would cause two problems. First, it would be impossible to tighten the chain by moving the rear wheel back, and, second, it looks ugly.

Once you have the main crankset and two chain guides in place, stand up your bike with the front wheel in place so that the seat tube (tube "B") is at 90 degrees to the ground. Your frame should look like the one in Figure 8-24, give or take a few feet. Note the placement of the two chain-guide cranksets, allowing the use of a single chain run.

Figure 8-24 Stand up the frame so that the seat tube is 90 degrees to the ground.

It is important to have the bike standing up so that the frame is in the proper position (with the seat tube at 90 degrees) so you can get the rear wheel in the correct position. Use a level or "plumb bob" to get it as close to 90 degrees as you can.

MAKING THE REAR END

The rear end of the bike is made from another set of forks and a short piece of 1.5-inch conduit or whatever you made the main frame out of. The forks will need to be stretched apart a little to take a rear wheel that has a wider axle. This is done by standing on one of the blades and pulling up on the other with a lot of brute force. When widening the fork legs, don't overdo it; just pull them apart enough so the rear wheel can fit into the dropouts.

The main part of the rear end is made by cutting the stem off the pair of widened forks and cutting the appropriate groove into the end of a piece of tubing, as shown in Figure 8-25. The length of the tube plus the forks should be the same length as the distance from the lowest bottom bracket to the dropouts on the front forks. This will put both wheels at an equal distance from the seat tube (tube "B").

Figure 8-25 Cut the end off a pair of widened forks and groove out the tube to fit.

Cut the correct length of tubing, groove it out, then weld the widened forks into the groove so the new piece of tubing is parallel to where the cut stem would have been. This is done by placing the fork blades and tube on a level surface, then welding them together. When complete, the new piece should look like the one shown in Figure 8-26. Notice the rounded groove at the end of the tube; this is done so it will mate with the lower bottom bracket.

Once you have the fork blades welded to the short tube, place the rear wheel (with inflated tire) into the dropouts, then position the assembly in the correct position on the bike as it stands up, with the seat tube (tube "B") at 90 degrees to the ground (see Figure 8-27). The end of the short tube is welded to the bottom bracket at the intersection of tubes "B" and "D." It is very important to have the bike standing up in the correct position on level ground at this stage, or your frame will

Figure 8-26 The fork blades are welded to the short piece of tubing.

Figure 8-27 Weld the rear of the frame in place with the frame standing up straight.

end up looking like the Leaning Tower of Pisa when complete, and you will end up doing the world's highest face plant!

WELDING THE PARTS

Welding this part of the frame is probably the hardest part of the entire building process, since you have to do it with the frame standing up and still get everything in perfect alignment. A good solid tack weld at the top of the joint will allow you to tweak the alignment before you put on the final welds. Make sure the rear wheel is 90 degrees to the ground, or parallel to the seat tube (tube "B"), as you weld the joint. If the wheel is not aligned properly, your bike will have odd handling or chain derailment problems—not a good thing at this height.

Once you get the main part of the rear frame welded solidly in place, carefully remove the wheels and lay the frame down on its side so the new rear part is pointing upward. Try to avoid hitting the rear of the frame, as it may be knocked out of alignment. The last two tubes that have to be added will be the seat stays (that's what they would be called on a normal bike). Two pieces of one-inch conduit or similar tubes are placed from each fork leg to the joint between the lowest step and the seat tube (tube "B"), as shown in Figure 8-28. These tubes complete the rear "triangle" and give the frame great strength and rigidity. If you ignore all of the tubing above the first step, the unit actually looks like a normal bicycle frame, just larger.

ADDING THE SEAT POST AND SEAT

We will now add the adjustable seat post and seat. Because the frame tubing is much too wide to add a seat post clamp directly to the end, a large washer is used as a bridge between the two. As shown in Figure 8-29, you will need a cut-off seat post clamp with an inch or so of the seat tube still attached, a large washer with the same diameter as the frame tubing and a center hole large enough for the seat post to fit into, and a seat post.

First, weld the small stub of the original seat tube to the inside hole of the washer. Make sure the hole is large enough to take the seat post. If the hole is too small, file it out with a round file. Once the seat post clamp is welded to the washer, weld the washer to the top of the main

Figure 8-28   The rear "triangle" of the frame is complete.

Figure 8-29 Parts used to make an adjustable seat post.

frame seat tube (tube "B") and grind the edges clean. Your adjustable seat clamp should look like the one shown in Figure 8-30 and be able to adjust to any height from zero to the full length of the seat post. If the seat post seems to get stuck in the washer area after you weld it all together, you may need to file out the hole a little more, as the heat may have distorted it.

ADDING THE CHAIN AND HANDLEBARS

Now that your Skycycle is almost complete, you only need to add a few more parts like the chain and handlebars before your first test ride. The long chain is made by joining many same-sized chains together until you have one long enough to cover the entire distance. Garage door openers have a chain that is the same size as BMX chain, so if you can find a source for these, you have a good supply. When joining chains, use only same-sized chains free of twists, kinks, and major rust. The key to the success of this project is the chain line working properly. Bent or misaligned sprockets or a rusty chain will cause your

Figure 8-30   Finished and working adjustable seat post clamp.

chain do derail in midflight, and you will have to jump off the bike, not a fun thing to happen while you're 10 feet in the air!

The proper length of chain will allow you to pull the rear wheel back slightly and eliminate the slack. If the chain is too long, it will be loose and floppy, and if it is too short, you will not be able to get it on the chain ring. Do not try to force a chain that is too short onto the chain ring or you may bend either the chain ring or, possibly, the frame. There is a correct length, and you will find it if you experiment. Your chain will run from the top crankset and be guided to the rear wheel by the underside of the two guide sprockets (see Figure 8-31).

GUIDE RINGS

In addition to a clean chain and aligned chain rings, there is one more method employed to keep your chain from falling off during a bumpy ride—guide rings. The chain is very long, so it has a tendency to flop around when you run over rough ground, and this could cause a derail-

Figure 8-31 The chain is guided to the rear wheel by the two guide sprockets.

ment. By placing a set of guide rings near the top and bottom, this problem is eliminated.

As shown in Figure 8-32, guide rings are made by cutting the threaded bodies off of screw-in eye hooks, found in many retail stores that have a hardware section. The hooks are placed on the steps closest the top crankset and bottom chain guide sprockets. This is done when the chain is in place and tight, to keep chain rubbing to a minimum, although you can expect a little rubbing as the chain bounces around. Be careful when you weld the guides in place, keeping your welding rod away from the chain. You may want to place some tape over the chain near the weld area to avoid welding the chain.

Before you brave your first test flight, it's a good idea to grind off the sharp edges left over on the lower guide sprockets where you cut off the crank arms. Just take the crank axle apart and grind the stub into a smooth round ball, as shown in Figure 8-33. This is also a good time to buff up the chrome and paint the chain guards as well.

Figure 8-32   Chain guide rings stop the chain from bouncing around over rough terrain.

Figure 8-33   Sharp edges are ground from the lower guide sprocket axles.

THE HANDLEBARS

As for the handlebars, any set will do. I prefer a nice wide "beach cruiser" style handlebar for stability and control, but this is totally up to you. A nice long gooseneck will also put the handlebars up a little higher so you don't have to hunch over like you do on a regular bike. Being as relaxed as possible while you fly on your Skycycle is a good idea. When your seat is properly adjusted and all the accessories are in place and tightened, you are ready to fly! Your Skycycle should stand straight and true like my 10-foot tall prototype in Figure 8-34. Note the placement of the chain guide rings.

RIDING TIPS

Riding the Skycycle is not difficult; in fact, it rides just like a regular bike. Getting up the nerve to climb to the top and start pedaling is difficult for the first time. Here's how it's done. Lean your bike against a telephone pole or sturdy post and have a friend hold the bike in place just to make you feel safer as you climb to the top. Once sitting in the pilot's seat, hold onto the pole, as shown in Figure 8-35.

You are now at the moment of truth. Either you will step on those pedals or slither back down the ladder in defeat! It always helps when your friend is laughing and calling you a sissy as you shake in your boots, trying to gather the nerve to take flight. This is your first ride, so test you bike out in a large, empty, wide-open parking lot and let your friends know that if you start yelling, they are to grab the bike so you can climb down. The big test will be to see if your chain will stay on without a derailment. I have ridden my Skycycle for miles in the street and never had a chain fall off, even when driving over large bumps.

So, 10 minutes have passed are you still hugging that telephone pole aren't you? Did you build your Skycycle too high? I told you so! When you do get the nerve to ride it, have a friend give you a little boost to help you get moving. Once you do a few laps around an empty parking lot, you will feel right at home in the pilot's seat. Although I would never ride the Skycycle without having at least one hand on the handlebars, I feel comfortable enough with the handling to stand on the pedals with one hand off the handlebars (see Figure 8-36). No kidding, it rides that good!

Landing the Skycycle is just the reverse operation of take off. Slowly roll up to a telephone pole or post with one hand ready to grab it as you ride by feathering the brakes. Once you have a secure grip on the pole, hit the brakes, and you have landed. Avoid coming in too fast and

Figure 8-34 The completed Skycycle is an awesome sight. Are you bold enough to ride it?

Figure 8-35 Letting go of the pole separates the brave from the meek!

Figure 8-36   Pulling a "one hander" is no problem once you feel comfortable with the Skycycle.

avoid getting so close that you hit the pole with your handlebars or pedals. Don't worry about the flaps, and always lower the landing gear!

The Skycycle draws more of a crowd than any other creation in this book, and for good reason. It's not very often that you see a person float by at 10 feet in the air with only two wheels and a few pipes underneath! Bystanders think the bike is really hard to ride because it has only two wheels, but you will see that it requires a lot less balance than a regular bike due to it's height.

Have you ever balanced a broom handle on the palm of your hand? It's fairly easy to do because it is so long that it falls over slowly. Try the same thing with a hammer or screwdriver. You will see that the longer the object, the easier it will be to balance. The same rule applies here.

The pilot seat of the Skycycle awaits you. Are you bold enough?

MULTIPLE RIDERS

The Mountain Tandem

CHOOSING AND ATTACHING THE FRAMES

Riding a tandem can be a lot of fun for you and a partner, but the cost of a new unit can be very high. However, with two old bike frames and a set of wheels, you can make your own for next to nothing.

Although you can join just about any two equal-sized bikes together, I decided to use two heavier mountain bike frames and wheels to build an off-road-capable machine—the Mountain Tandem. The two frames to be used in this project should have similar size and geometry. The closer the two frames are, the better the final product will be.

As shown in Figure 9-1, the two frames I used are identical in every way except for the labels. If you plan to ride off road, then try to find heavier built frames with oversized tubing like the ones I have chosen. In addition to the frames, you will also need a good set of front forks and two identical cranksets. The cranksets must have the same size chain rings in order to synchronize them with the connecting chain. If one of the chain rings had a different tooth count than the other, a collision of feet between riders would eventually occur.

Choose which frames will be used for the front and the back. If they are both in the same condition, it makes no difference, but if there is some damage to the front or rear of one of the frames, then plan your project so that the best parts of the frames are used and the worst parts are removed.

Begin cutting the seat stays and chain stays from the frame to be used as the front, as shown in Figure 9-2. The rear frame will have the head tube and down tube removed as well. Cut the head tube as close to the joints as you can in order to keep the top tube as long as possible. The down tube from the rear frame is then used as the boom to

Figure 9-1   Choose two frames of similar size and geometry for the Mountain Tandem.

Figure 9-2   Frames are cut and rearranged to form the basic Mountain Tandem frame.

join the two frames together at the bottom brackets. So far you haven't needed any extra tubing!

Lay the frame out on a level surface, as shown in Figure 9-2. The goal is to joint the two frames together so that both seat tubes are at the same angle and both top tubes form a straight line. To achieve this, make sure that the boom (lower joining tube) is the same length as the top tube on the rear frame. Don't forget about the half inch or so of length that will be lost when you groove out the ends of the tubes to make a good-fitting joint, especially on the boom tube because it mates with the bottom brackets at both ends.

Once you have the tubes cut and ground for proper fitting, tack weld the two frames together so that they look like the one in Figure 9-3. If

Figure 9-3   The two frames joined together with the proper geometry and angles.

you cut the boom to the correct length (same as the rear top tube), then your frame should look balanced, with both seat tubes at the same angle and both top tubes forming a single straight line.

The hardest part of joining the frames is achieving the correct alignment. As you weld one side of a joint, heat will pull the two frames out of alignment, and you will need to apply some force to get them straight again. This is the hardest part of the building process, and if you take your time and switch sides as you weld, it can be done without too much trouble.

To check the alignment, hold your frame up by the head tube while the rear dropouts are on the ground and look down the length. The two top tubes should form a single straight line when they are welded in place. If your frames end up out of alignment, use some muscle strength to get them straight.

THE TRUSS TUBE AND TENSIONER

Although the two frames are fairly strong now, the addition of a truss to form a triangular structure between the two seat tubes will greatly enhance the overall strength of the final product. As shown in Figure 9-4, this truss tube is added from the top front corner to the rear bottom corner between the two seat tubes. You could put this tube in the other way as well, but it just looks nicer this way, since it follows the same angle as the down tube on the front frame. The truss tube does not have to be a large, heavy tube. I used a one-inch piece of conduit as the truss, but you could even get away with a ¾-inch piece if you had some lying around.

Now that your frame is complete, there is only one more weld to make—the chain tensioner. A rear derailleur is welded to the center of the boom (see Figure 9-5). This will be used to pick up the slack in the return chain that runs between the two equal sized chain rings. There is no other way to adjust this chain besides pulling up on the bottom, and making a chain the exact size would be almost impossible unless the two chain rings had exactly the correct space between them. On a professionally built tandem, the front bottom bracket is mounted in an offset ring inside a larger bottom bracket, allowing it to move forward or backward about an inch. Making this type of thing would require the machining of a lot of material.

This derailleur chain tensioner system works quite well, and I have never lost the chain yet, even when driving off road on very rough trails. A derailleur with a good return spring works best. You can also adjust the limit screw on the derailleur to move it side to side slightly

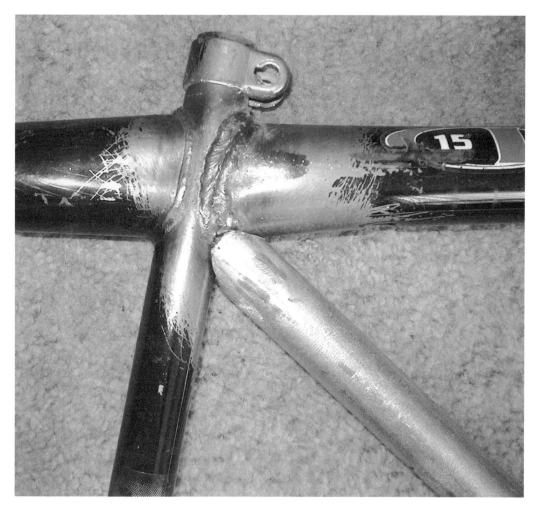

Figure 9-4 The truss added between the two seat tubes gives the frame extra strength.

in order to align it perfectly with the two chain rings. This will reduce noise and friction.

ATTACHING THE GOOSENECK

Your passenger (stoker) will need something instead of the back of your head to hold onto while he or she rides, so the end of a gooseneck is welded to the captain's seat post (see Figure 9-6). The cut-off end of the gooseneck is welded so it is about an inch below the start of the taper where the seat will be mounted to the post. If you put the goose-

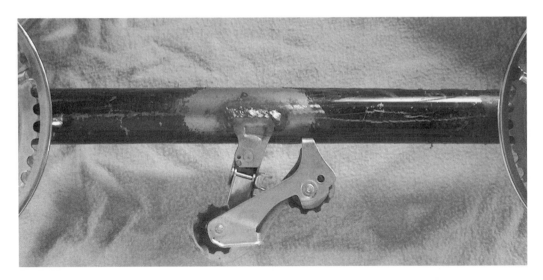

Figure 9-5   A derailleur is welded to the boom in order to pick up the chain slack.

Figure 9-6   The stoker's handlebars are mounted to the captain's seat post.

neck up too far on the seat post, the seat will hit the handlebars, or the captain's legs will hit the handlebars.

Tack weld the gooseneck to the seat post while the handlebars are mounted and the seat post is connected to the bike. This way, you can ensure that the handlebars will not be mounted at an angle when you are done welding. Although the handlebars will be able to move up and down in the gooseneck, they should be aligned perfectly horizontal or 90 degrees from the seat post. When you have the gooseneck tacked and aligned to the seat post, weld it all the way around, making sure that it is securely connected.

Now you can put all the components onto the Mountain Tandem and see how it works. Your final product should look like the one in Figure 9-7 if all went well. I decided to use a tall set of handlebars from a freestyle bike rather than the straight type you find on a standard mountain bike. This way, I would not have to hunch over while I was

Figure 9-7 The Mountain Tandem is ready for the first test run.

riding. Also, the stoker's handlebars are a little wider so they will not rub against any part of the captain's body while riding.

CRANKSETS AND CHAIN MANAGEMENT

When you connect the chain between the two equal-sized chain rings, make sure both cranksets are in the same position so both will be synchronized while pedaling. The smaller (inner) chain ring is used for the connecting chain so that the larger ring on the back crankset can be fed to the rear wheel. If you did this the other way, you would always be in low gear.

Depending on the style of crankset and thickness of chain you are using, there may have been some rubbing where the connecting chain meets up with the drive chain at the back. As you can see in Figure 9-8, they both have to share the rear crankset, and depending on how close the two chain rings are together, they could rub. There are a few solutions to this problem if it occurs. With the crankset that I used, there was only a small amount of chain rubbing, and because the chain

Figure 9-8 The connecting chain and drive chain must not rub together.

rings were made of steel, I just pried them apart carefully with a flat-head screwdriver. This is accomplished by slipping the blade of the screwdriver between the chain rings and prying them slightly apart about every two inches until you have done the entire circumference. If you do this carefully, the two rings will be moved apart without any warping. This method will only work on a steel crankset, not an aluminum one.

The other method used to separate the chain rings is to remove the bolts or rivets that hold them together and add a washer in between the two rings. Since the cheaper cranksets are pop-riveted together, you will have to grind the heads off the rivets than replace them with the proper-sized nut and bolt. This method will work on any style crankset, and will ensure that there is no warping in the finished bike.

Once you have your chain management problems solved, you can give the bike a test ride. In addition to the wider turning circle, you won't notice much difference between this bike and a "regular" bike. Finding a long enough shifter cable may be a bit of a challenge (see Figure 9-9). The cable from the rear derailleur to the front shifter on the finished Mountain Tandem has to make a fairly long journey. If you can't find a cable long enough, have a bike shop cut you off a long enough piece from a roll.

Brakes are important as well, since you will be carrying almost twice the weight as on a normal bicycle. The front brake will do most of the work, so make sure it is well adjusted and set up properly. Due to the extra length of cable needed to run to the rear brake, the overall effectiveness will be reduced because cables add friction and can stretch when they are pulled. The brakes to use are ones that are mounted directly to the fork legs.

RIDING THE MOUNTAIN TANDEM

Riding the Mountain Tandem is easy once you and your passenger learn to do things in sync. When mounting the bike, the captain should get on first, standing with both feet on the ground and a good grip on the handlebars. The stoker then gets on the bike, and both riders then push down on the same side of their respective pedals. The stoker should always wait for the captain to get a firm hold of the bike before climbing on, or both may end up lying on the ground.

While riding, there are no special rules to follow since the captain has full control. The stoker should avoid any rocking back and forth or large movements while the captain is riding with one hand, since this could upset the steering, causing an unexpected sharp turn.

Figure 9-9 The completed and painted Mountain Tandem ready to roll.

While making a sharp, fast corner, the natural tendency is for both riders to lean, but the stoker should not overdo it. Stopping the bike is done in the reverse order from starting. The captain gets both feet firmly planted on the ground and then tells the stoker it's OK to dismount. If the captain is not ready and the stoker climbs off the bike, both riders will be seeing a close-up view of the ground as the bike falls over. Yes, I've seen this happen. If the captain is allowed to call the shots, everything usually works out all right. Once you lean to ride in sync (see Figure 9-10), the Mountain Tandem can be fun for all ages.

I hope that you enjoy the Mountain Tandem! Everyone likes to have a ride on the bike, so I am currently searching for a few more same style frames so I can extend the unit into a three- or four-seater just for fun. But until then, the next project will show you how to add an

Figure 9-10 Ma and Pa (Lillian and Tom) Graham take the Mountain Tandem for a cruise.

extra passenger to the bike with little effort and a few scrap parts.

The Detachable Tandem

The Mountain Tandem is great fun for older kids and adults, but if younger kids want to come along for a ride, they will not be able to reach the pedals due to the size of the frame. You could have the kids tag along on their own pint-sized bicycles, but this also may be a problem because of their lack of speed or ability to make the long journey.

The Detachable Tandem is a solution for the smaller bicycle enthusiast in the family, allowing them to "tag" along yet still participate by doing some of the pedaling. This simple "half-bike" attaches to a special x/y joint on the main bicycle's seat post, allowing the unit to move up and down, left and right, but not allowing the unit to tilt side to side and fall over. Because of this, the rider on the detachable can feel like they are riding a regular bicycle, yet there is no chance of them falling over (unless, of course, the main bicycle falls over).

CHOOSING A WHEEL

This Detachable Tandem can be built in a night or two from common parts you most likely already have in your "scrap pile." The main part of the project will require a kid's bicycle frame and rear wheel to fit like the one in Figure 9-11. There are many different sizes of kid's bikes with wheels from 12 inches to 24 inches but, for best results, use a wheel no smaller than 20 inches. If you choose a wheel smaller than 20 inches for the Detachable Tandem, then the child riding it would never be able to contribute any pedaling power due to the incredible speed at which they would have to rotate the cranks in order to match your pedaling speed.

With a 20-inch wheel on the rear of the Detachable Tandem, it would be quite easy for a child to pull his or her own weight as part of the riding team, although you can't expect the young ones to keep up the pace when you are blasting around at more than 18 miles (30 kilometers) per hour. If you do happen to "outrun" the usable gear range of the De-

Figure 9-11 A kid's bike frame and 20-inch rear wheel are best suited for the Detachable Tandem.

tachable, the rider just stops pedaling and glides until the speed returns to normal, because the pedals are fully independent of yours.

THE SWIVEL JOINT

Once you have your donor frame and wheel, put it away for now. First, we will create the most important part of this project—the swivel joint. This swivel is the key to this project because it allows the detachable to move up and down as well as left and right, but not sideways. This may seem a little confusing at first, but think of the reasons behind this design.

If the Detachable could not move up and down, it would either be lifted off the ground if you rode over a large, steep hill or snap right off at the joint. You could not have all three wheels on the ground at the same time unless one of them could move up and down to conform to the slope in the grade. The Detachable also has to move left and right so the main bike can steer. This system works the same way as a truck pulling a trailer.

Now, you may be wondering, why not just use a ball joint or something similar to a trailer hitch? At first, this would seem logical because it would allow any type of motion at the joint, including up, down, left, and right. The problem is that it would also allow the Detachable to move from side to side or roll, which means that it would fall over even if the main bike were standing straight. Because the Detachable does not have front wheel or steering, it cannot balance independently of the main bike and would fall over and be dragged along. Imagine a truck pulling a trailer with only one wheel!

Now that you understand the rocket science involved in this project, let's start making it work. You will need a decent front hub from any sized wheel. This can be removed the nice way by undoing all the spokes, or the fast way by cutting them out like I did in Figure 9-12. This hub will form the part of the swivel joint that allows the Detachable to move up and down. There will be parts welded to this hub, so make sure that it is made of steel, not aluminum.

The other two pieces that make up the swivel joint can be cut from a short piece of one-inch thin-walled electrical conduit or similar-sized tubing (see Figure 9-13). You will need two pieces that are two inches in length. The tube chosen must be able to fit over the seat post on the bike you plan to pull the Detachable with. The tube should fit fairly snug, but not so tight that is has to be put on with force. It should be tight enough so that there is not a lot of slop or gap as well. One-inch electrical conduit fits just right over a seat post, but if you do not have

Figure 9-12   A front hub in good condition is cut from the wheel.

any, there are many other types of tubing, including the seat tube cut from an old frame, that would also fit.

Grind a groove out of the end of one of the two-inch pieces of tubing so that it can be welded to the shaft that forms the front hub. The groove should be about as deep as half the diameter of the hub shaft, as shown in Figure 9-14, so the two can be welded together easily. When you are welding the two parts together, watch your heat and take care not to burn a hole into the hub, or you may end up with a seized axle. Also, this tube should be connected so that it forms a 90-degree angle with the hub shaft.

Once the piece of tubing is welded to the hub, you can then groove out the other end of the tube so the second piece of tubing can be welded to it in a "T" shape, as shown in Figure 9-15. Because this is the tube that must slide over the top of the main bike's seat post, take care not to burn a hole through it or warp it badly from too much welding

Figure 9-13 The two small pieces of tubing that form the rest of the swivel joint.

heat. If you do burn a hole through this tube, you will have to use a hand file to remove any imperfections on the inside.

When this tube is mounted around the main bike's seat post, it forms the joint that allows the Detachable to move left and right so the two bikes can steer like truck and trailer. It is important that this tube is welded into position so that it is 90 degrees from the hub. The hub will form the pivot point for up and down motion, and the seat post tube will form the pivot point for left and right motion. If the two joints are not welded at 90 degrees to each other, the Detachable will end up leaning to one side in the final product.

ADDING THE FORKS

Once the swivel joint has been welded together, place it into the dropouts of a pair of forks that will be used to create the "hitch" for the Detachable (see Figure 9-16). Any set of forks will work fine as long as

Figure 9-14 The end of the two-inch tube is grooved out to mate with the hub shaft.

Figure 9-15 The second piece of tubing is welded to form a "T" shape.

the hub will fit into the dropouts. I used a very small pair of forks from a kid's bike with a 12-inch wheel in my design, but any length will do the job. Also, notice the special "hooked" washers that hold the axle to the dropouts, as shown in Figure 9-16. These will stop your hub from falling out of the dropouts if the bolts become loose.

If your hub comes loose from the forks, your Detachable will be let go and end up in a nose dive, so make sure that this never happens. You may even want to tack weld the nuts right to the dropouts since there will be no need to remove the hub once the project is completed.

As shown in Figure 9-17, the swivel joint and forks are placed on the seat post of the main bike so you can test to make sure it all works properly.

ADDING THE SEAT POST

The mating seat post must be raised at least as high as the length of the tube that is placed in it so that the seat can be placed back on the tapered end if the seat post. If you built your swivel and hitch correctly, you should be able to move it in any direction left and right as well as up and down but not be able to twist it. Now you can see the true magic in this design—it allows the Detachable Tandem to be pulled

Figure 9-16   The "hitch" will be created from a set of forks that hold the hub.

Figure 9-17   The completed hitch and swivel connected to the seat post for testing.

along like a trailer yet not fall over no matter what the passenger does. Of course, if the passenger has a lot of spunk, he or she may actually push you along for the ride!

If there is excessive slope between the seat post and tube, or if you are worried about scratching the seat post, then you may want to place a small "bearing" surface in between the two. For this job, a small piece of cardboard or thin copper foil would work nicely. If there is not enough room for anything to fill the gap, then just add a little oil or grease to the joint before you ride. There will be so little movement here that this will most likely be unnecessary.

MAKING THE FRAME

Now you need to work on the actual Detachable Tandem frame. Start by cutting off the top tube and down tube ahead of the seat tube, as shown in Figure 9-18.

You can grind off the stubs left over on the seat tube after they are

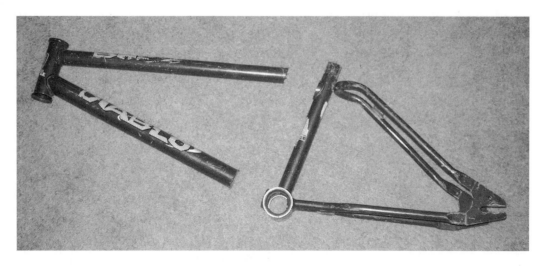

Figure 9-18   The top tube and down tube are removed from the donor frame.

cut from the frame. The goal is to mount a length of tube from the joint where the top tube used to connect with the seat tube to the forks used as the hitch. The length of this tube is determined by how far back from the main bike's wheel the cranks on the Detachable Tandem will need to be. Since a typical crank arm is about seven inches long, any distance between 10 and 12 inches from the back of the tire to the center of the bottom bracket on the Detachable will be safe.

I set my freshly chopped frame up on a block (see Figure 9-19) so there would be adequate clearance not only between the cranks and the rear wheel of the main bike, but also the cranks and the ground. Once this is determined, take the measurement from the top of the seat tube on the Detachable to the top of the forks that make up the hitch. The tube is then cut to this length and placed between the two pieces to make sure it will not rub on the top of the main bike's rear tire.

The tube that makes up the main hitch is made from a length of 1.5-inch muffler pipe or similar sized tubing. You could probably get away with one-inch electrical conduit, but this would be pushing the limit for a rider over 100 lbs. If you cannot find any 1.5-inch round tubing, a square tube could also be used and this is pretty common material.

When you are satisfied with the placement and length of the main hitch tube, weld it directly to the hitch forks, as shown in Figure 9-20. Weld the two parts so they are aligned perfectly, as if the new tube was just an extension of the original fork stem. Since the original fork stem is not used and would only end up hidden inside the hitch tube, it can

Figure 9-19 The Detachable Tandem must have adequate pedal clearances.

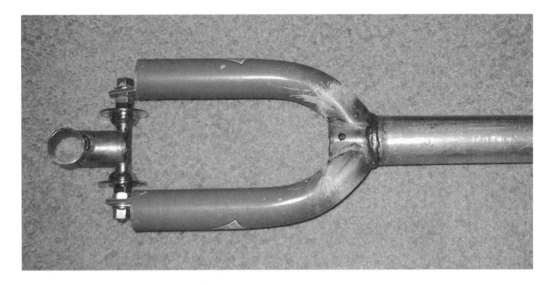

Figure 9-20 The main hitch tube is welded to the forks at the crown.

be removed and added to your parts bin for later use in some other project.

ADDING STRENGTH

Your frame is almost complete. To give the Detachable Tandem that added strength, one of the spare tubes cut from the front end of the frame is added so it forms a triangle between the bottom bracket, seat tube, and hitch tube. The length of this tube is not critical as long as it forms a triangle, so any length between 12 and 18 inches would be just fine. As shown in Figure 9-21, this new tube is cut and grooved out to fit into its new place on the frame. The tube will be welded in the place where the original down tube connected to the bottom bracket and it will extend upward to the main hitch tube.

Then, cut, groove, and weld a steel gooseneck to the top of the hitch (see Figure 9-22). The best place to weld the gooseneck is at approximately the same distance from the seat tube that it was originally on the bike. This way, the passenger will feel right at home. In my design, this distance was right over the top of the extra tube added for

Figure 9-21   An extra tube will be added to form a triangular structure in the frame.

Figure 9-22   A steel gooseneck is welded to the top of the hitch tube.

strength and this made for a nice-looking final design.

ADDING ACCESSORIES

Now that all welding is complete, you can add the accessories such as the seat, handlebars, and crankset. Your final project should look like the one shown in Figure 9-23, with adequate clearance between the hitch tube and main bike's rear tire. There is nothing special about any of the accessories, since the back end of the Detachable is pretty much in its original working state. A larger chain ring (with 52 teeth) was added to the Detachable Tandem so the rider would be able to keep up with the main bike. The chain ring that originally came with the kid's bike I used had so few teeth that the rider would have to pedal like crazy just to keep up with the slow pace of the main bike. Kids'

Figure 9-23 The completed Detachable Tandem with a larger front chain ring added.

bikes are not really made for any real speed, but since this one is being pulled along in a safe manner, it is OK to increase the gear ratio to something more usable for distance riding.

RIDING TIPS AND TRICKS

Riding a bicycle with the Detachable Tandem connected is very easy. In fact, you can barely tell it's even there! The same rules should be followed as those for a stoker (passenger) on a regular tandem. Wait for the captain (driver) to get on the bike before the Detachable is mounted, and wait for the captain to give the signal to all passengers that it is safe to dismount. Although a kid's weight is not enough to topple you and your bike if they climbed aboard before you where ready, it doesn't hurt to have control of your ship.

The Detachable Tandem not only hooks up to a regular bike, but can also be connected to a Tandem, as shown in Figure 9-24. Now we have a two and a half horsepower vehicle!

This little add-on unit works great and is a lot of fun for the kids, since they can now feel like part of the team because they are doing some of the work as you ride. There is also no worry that they may

Figure 9-24 Tyler (riding the Detachable), Tracy, and Brad try some bicycle teamwork.

wander off into danger because they have to follow your lead.

Now that you've seen how to *add* one wheel and a rider to a tandem, in the next chapter you will see some of the wild and whacky contraptions you can create by *using* only one wheel!

10

ONE-WHEELED WONDERS

The Giraffe Unicycle

Here's a fun project that will help hone your Jedi reflexes—a five-foot-tall unicycle. Unicycles are challenging to learn to ride, but with a day or two of practice anyone can do it. If you have never ridden a unicycle before, you may want to start off with a smaller, direct-drive unicycle (the ones with the crank arms attached directly to the wheel hub). A smaller unicycle is easier to get up on when you are just starting out and, yes, a lot closer to the ground when you are falling off!

If you think you can handle this monster, then start digging in your scrap pile for a few common bicycle parts. You will need a decent rear wheel with a steel hub (any size will work), an old frame to cut a bottom bracket and seat tube from, and some short lengths of electrical conduit or similar-size round tubing. The most important part of this unicycle, besides the skill to ride it, is the fixed sprocket hub.

A unicycle must be pedaled in both directions in order to make corrections for balance and to give feedback to the rider. On a small direct-drive unicycle, this is no problem because the crank arms are directly attached to the wheel hub. But on a tall unicycle, we must do something different since the cranks are above the wheel. To simulate the feel and response of the direct drive unicycle, we will be welding a chain ring directly to the wheel hub and running a chain back to our crankset. This way, we can pedal in both directions and make quick balance decisions, just like it is done on the smaller unicycle.

REMOVING THE FREEWHEEL CLUSTER AND BEARINGS

Let's start by removing whatever type of freewheel cluster is attached to your hub. Again, make sure the hub is steel not aluminum, since we will need to be able to weld the chain ring to it. To remove the freewheel, you will need a hammer, punch, and plastic container. The plastic container is set under the hub to catch the 20 million small ball bearings that will be falling out as we remove the freewheel from the hub. Place you wheel on top the container, as shown in Figure 10-1, and place a punch into the groove or hole on the top ring of the freewheel.

Tap the hammer so that the ring turns in the clockwise direction. Once it is moved slightly, it will unscrew easily unless it is bent or damaged in some way. If the wheel you are using is a 20-inch BMX-style freewheel with one gear, then you will need to tap the ring in the counterclockwise direction because this unit will unscrew directly from the hub in one piece.

Once the top ring has been removed, lift the gear cluster off the hub, making sure that the plastic container is under the hub. A freewheel contains more ball bearings than any other part of a bicycle, as you will soon see, and they will roll everywhere if not caught in a container. There are two layers of bearings, and you can see the top layer in Figure 10-2. The bottom layer has three times as many as the top. If you ever have to put a freewheel back together, you have to put a thick layer of grease on the bearings so they stick in place, then carefully place the hub back into the freewheel; but if you value your sanity, don't bother trying this.

Figure 10-1 Removing the freewheel from the steel hub.

Once you have the gear cluster and bearings out of the way, remove the small ring that holds the pawls (the things that making a clicking sound when you pedal backwards) from the remaining part of the freewheel. Now place a pipe wrench or really big vice grip onto the remaining part of the freewheel and unscrew it in the counterclockwise direction. This piece may take a lot of force to remove since it has been tightened by the force of pedaling for as long as it has been on a working bicycle. Placing the rim on the ground and pulling up on the pipe wrench is the easiest way to get good leverage, but don't slip and give yourself a black eye! Once all of the freewheel has been removed from the hub, it will look like the one in Figure 10-3.

ATTACHING THE CHAIN RING

OK, now the fun part. You will need to find a one-piece crankset and remove the chain ring from it. If you don't know what I'm talking

Figure 10-2 A freewheel contains many small ball bearings.

about, refer back to Chapter 5. I call this the fun part because you must make the small hole in the middle of the chain ring (where the crank arm went through) large enough to fit over the threaded part of your wheel's hub, and it must be accurate or the chain will be loose.

There are two ways to do this that I can think of—with a cutting torch or by hand filing out the hole with a round file. In the true Atomic Zombie spirit, I choose to do it by hand (I don't have a cutting torch). Hand filing the hole took a few hours, but allowed me to make small corrections to the hole as I went along. If you cut the hole cut with a torch, you may end up hand filing it to get it perfectly round in the end, anyway. Figure 10-4 shows the chain ring with the hole ready to fit over the hub's threaded end.

When the chain ring is welded to the hub, it must be as straight as possible and centered or your chain will be falling off and so will you. If the chain ring is not centered, the chain will go from loose to tight each

Figure 10-3   A naked hub, stripped of the freewheel.

time the wheel makes a revolution and this will derail the chain. Place your wheel's axle into the vice (between two blocks of wood) so the wheel can be spun around. Welding the chain ring to the hub is going to take some patience, and it will take several attempts to get it aligned, unless you get lucky on the first try.

The best way to start is by making only a small tack weld on one side of the chain ring to secure it to the hub. At this point, you can spin the wheel around and tap the ring around with a hammer to get it straight. If the ring seems to move up and down, then you will have to break the tack weld and try again, since there is no other way to align it. I had to do this three times before I managed to get it straight enough, but it didn't really take that much effort.

When you do get the chain ring set up nicely, add another tack weld to the other side, and repeat the spinning and aligning procedure. You will not have to break the weld again once the chain ring is centered, but you will have to make small corrections to the warpage of the chain ring caused by welding heat. Try to make small welds on oppo-

Figure 10-4   Hours of hand filing makes a fairly round hole.

site sides as you weld the chain ring to the hub. This will cut down on heat distortion. If all goes well, leave the wheel in the vice, and grind the weld as the wheel spins. This will make a nice-looking completed weld, like the one shown in Figure 10-5.

CUTTING THE DONOR FRAME

OK, now you can relax because the rest of this project is a breeze, at least until you try to ride this wild animal! You will need to find a donor frame to cut the bottom bracket and seat tube from. If your frame is made of all the same size tubing like the one I used, then you can cut it so either the seat tube or down tube remains on the bottom bracket; but if the down tube is larger than the seat tube, use the seat tube. We want to be able to insert the seat post back into the tube. This is why we need either the seat tube or same-diameter tube. Cut the frame as shown in Figure 10-6 so that either the seat tube or same diameter main tube remains on the bottom bracket.

Figure 10-5 Chain ring and hub welded together.

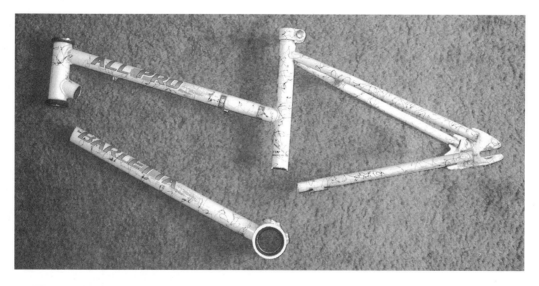

Figure 10-6 The bottom bracket and seat tube will be removed from the donor frame.

Once you have the bottom bracket and seat tube out of the frame, the rest of the project is fairly simple, and you can make any changes to the height of the unicycle as you see fit. If you plan to be a professional circus clown or just want to sample the great taste of pavement, then feel free to make your unicycle 15 feet tall. This design will permit it with little very change. You will need a few pieces of one-inch conduit or tubing from another bike frame to make the rest of the unicycle. The best thing to do is lay out your rough design on the floor, as shown in Figure 10-7. This method will help you find the right lengths of tubing needed. The only thing that really matters here is that the final width of the fork legs match your chosen wheel's axle, and this determines the length of the horizontal tube that joins the two fork legs to the top frame. In my design, I used one-inch conduit for the for legs, and 1½-inch conduit for the fork top and the small tube that joins the bottom bracket to the fork top.

The dropouts that I used were cut from the donor frame that the bottom bracket was taken from, but anything would work, including the ends of an old pair of forks.

WELDING THE MAIN BOOM AND FORK LEGS

Once you have decided on the layout of your unicycle, weld the main boom (the tube that determines the height) to the bottom bracket, as

Figure 10-7 Laying out the rough design and deciding how tall you want the unicycle to be.

shown in Figure 10-8. Remember that not only can you build it too high, but also too low. If the main boom is too short (less than seven inches), the crank arm will hit it as it turns, and your unicycle journey will last approximately two feet before you face plant. If you do decide to build your unicycle to some ridiculous height, then use at least a 1½-inch tube for the main boom to give it that extra strength. Try to get all tubes welded as straight as possible, as this will be the deciding factor in keeping your chain from falling off.

Once you main boom is welded in place, you can weld the top of the forks to it. This is the piece that determines how wide your forks are, and should be at least as wide as your wheel's hub plus the width of the two fork legs. This tube must be welded at 90 degrees to the main boom or else the wheel will be out of alignment, and this will only make the unicycle harder to ride, something you won't need. If the tube is not welded on straight, grind off the weld and do it again because this part counts.

The next parts to be welded are the fork legs. Make sure both legs are exactly the same length before you weld them, or it will be hard to correct them afterward. The best way to get things aligned properly is to tack weld the outside of the fork legs to the fork top tubes (see Figure 10-9) so you can work them with a hammer if necessary. Make sure the two legs are far enough apart so that the wheel can be mounted once the dropouts are in place.

Figure 10-8 Main boom welded to the bottom bracket.

Figure 10-9   Fork legs tack welded to the top fork tube for easy alignment.

Once again, take your time and get the two legs as straight as possible to avoid balance problems or chain misalignments. As you make small welds, check your work from all angles, making any necessary adjustments with your good friend the hammer. A professional frame builder has a jig to hold everything in place while welding but, with patience, your work will be just as good. Look down the length of the frame, as shown in Figure 10-10, and check the fork legs for alignment. It's a lot easier to move them around before you put in the final welds.

MOUNTING THE WHEEL

Your frame is almost complete. To mount the wheel, we need a pair of dropouts that will allow the axle to move up and down for chain tensioning. The dropouts must allow at least a half inch of travel in order to get our chain tight. Any set of dropouts from the rear of a frame will do the job, or you can make your own my cutting a slot out of a piece of scrap plate steel. Whatever you do, make sure both dropouts are the same size and shape, especially if they were cut from a frame using a grinder. To make them exactly the same, clamp them both together in a vice so the slots are aligned, and grind them together until they are of equal size and shape.

Welding the dropouts to the fork legs is easy. Just mount them the axle and place the wheel in between the fork legs so everything is in

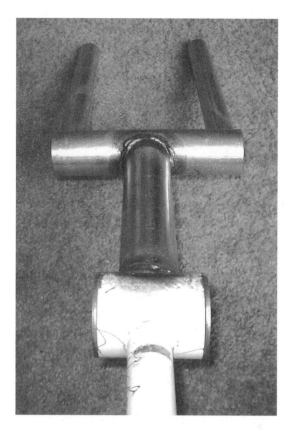

Figure 10-10   Make sure the fork legs are aligned before the final welding is done.

the correct place, then tack weld the dropouts to the fork legs. Once you are sure the wheel is aligned, remove the axle bolts and finish the welding of the dropouts. If you leave the wheel in place when you are welding the dropouts, welding spatter may damage the axle threads or melt through the tire. The finished dropouts should look like the ones in Figure 10-11—clean and ground. I filled in the remaining end of the tube by placing a small bolt in the end and welding over it. Once ground, it will look like one solid piece.

ADDING THE SEAT AND CHAIN

All that you need now is a seat. To be able to adjust the seat post, a seat post clamp will be needed. If you cut the donor frame leaving the seat tube in place, then you can skip this step. To make an adjustable seat post, cut a slot out of the top of the tube with a grinder or hacksaw

Figure 10-11 Dropouts welded to the fork legs.

(see Figure 10-12) and place a seat post clamp from another frame back over it. You're done!

Are you getting nervous yet? You're only minutes away from your first test ride (and fall)! Once your seat post clamp is ready, it will look like the one in Figure 10-13. Make sure the seat post can be tightened fairly snug as there will be a lot of twisting force on it as you ride. It may be worth your while to get a real unicycle seat for your unit, as these help make your ride a little easier. A true unicycle seat is rounded to allow you to sit in it rather than on it. This helps keep the unicycle in between your legs as it moves around under you during your balancing act. You could just use a regular bicycle seat but, as a beginner, a real unicycle seat will make your life a lot easier.

Go ahead and add the chain and seat to your creation and see how it all fits together. The chain should be adjusted so that it is fairly tight,

Figure 10-12   A slot is cut in the tube to allow the seat post to adjust.

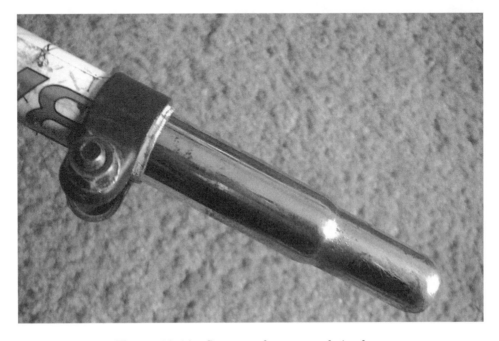

Figure 10-13   Seat post clamp securely in place.

Figure 10-14 The completed giraffe unicycle ready for test riding (or crashing).

but not so tight that pedaling is met with a lot of resistance. Your seat should be at the correct height for your inseam; possibly an inch lower, but definitely not an inch too high.

When the chain is connected and adjusted, turn the unicycle upside down and give the pedals a few turns. If you took your time with the alignment of the chain ring to the hub, everything should be moving smoothly, with the chain remaining at the same tension throughout the pedal rotations. If not, well you know what to do—go fix it.

My completed unit, shown in Figure 10-14, has a real unicycle seat attached. Because I am not a skilled circus clown and have never ridden a tall unicycle, I opted for the real unicycle seat in order to get all the advantage I could when learning to ride this thing for the book picture. After all, I don't expect you to build this thing without me proving to you that it can actually be ridden, and since I can only manage a few feet on a normal unicycle, this was a real challenge.

I chickened out of my first test ride and decided to paint the unicycle first so it would look better when it landed on top of me, but now that

Figure 10-15 Getting ready for the launch.

the paint has dried, it's show time. I will not pretend to be an expert unicyclist, but will share my experiences with you in hopes that it may help you learn to ride this beast like I did (see Figures 10-15 and 10-16 for the proof).

The best way to get on the unicycle is to climb up a ladder or have two tall friends help you balance. When you are falling, try to catch the unicycle so it doesn't crash on the pavement because this is hard on the seat (or the back of your head). Flail your arms wildly to help you balance, even though you may look like you're trying to fly. When you do manage to ride farther than 12 inches, use you hips in a swinging motion to steer the unicycle if you plan on going somewhere else besides in a straight line. Other than that, just practice. If a bear can learn to ride one of these things, you should be able to as well, shouldn't you?

Learning how to crash gracefully is also an important part of unicycling, as I found out. In fact, there isn't really a graceful way to get off

Figure 10-16   Swing your arms wildly. It helps you maintain balance.

a five-foot-tall one-wheeled bike that I know of. You just fall off. Like they say in the airline industry, "any landing you can walk away from is a good one," and that also applies here. If you can get off your unicycle, and catch it before it bounces off the road, you're doing great, and I salute you.

This project has been a lot of fun, and worth the effort put into it. The best part of unicycling is not riding the machine but watching your buddies try, too! Good luck on your unicycle journey.

The Wild Bull

The Wild Bull is truly a "one-wheeled wonder." In fact, it's a wonder that it can be ridden at all! When I first sketched out the design for this crazy contraption on paper, I was told that it would never work, so, naturally, I headed right out to the garage to build it. The idea behind the bike is simple.

Because the rear sprocket is welded directly to the rear hub and the front sprocket is about the same size as the rear, this will give the bike a one-to-one direct drive just like a normal unicycle. Of course, it does have a more recumbent position and a set of handlebars, but, really, how much different than a unicycle could it be to ride?

After building the Wild Bull, I wasn't quite sure if it was going to make it into this book since I have a strict rule that all bikes should actually function, so I went out to the street to give it a go. Initially, I just kept flopping and flopping onto the pavement like a freshly caught fish in the bottom of a boat, and it seemed to be going nowhere. Two hours went by and, under the dim glow of the streetlight, something amazing happened. I tried a new technique and rode the crazy thing for over 100 feet nonstop. Actually, Devon suggested that I try riding with only one hand, and since I was not really getting anywhere, I decided to give it a try just for fun.

The reason it worked so much better with one hand is because this allows you to swing some weight around with your arm as you fall sideways. Keeping the bike upright was an easy thing to learn, but staying vertical from side to side was a challenge. In fact, it was this crazy-looking, one-handed arm flailing that led me to the name "The Wild Bull." If you have ever seen those wild-bull riding competitions on TV, then you will probably remember the loose, one-handed technique they use to stay on the bull. Yes, it looks just like that when you ride this thing. Once you find that "sweet spot" where your body's center of gravity is directly over the frame, you can ride just about as long as you want, or until you hit a pothole in the road. Learning to stay on the Wild Bull was a lot of fun, and it only took about two hours of practice before I could ride half a block or so. Do you want to amaze your friends with your impossible balance? Then go dig up an old bike from the scrap pile and build the Wild Bull.

REMOVING THE FREEWHEEL

The most important part of the Wild Bull is the fixed rear sprocket. Just like the Giraffe Unicycle project, we need to weld a rear sprocket directly to the rear hub for that direct-drive effect. This time, we will just gut the insides of the rear cluster and weld it to the hub so it can't move. This will cut down on the amount of work necessary, and also allow for the changing of gear ratios if you so choose.

Start by removing the retainer ring on the top of the freewheel in the clockwise direction. Make sure you have a small container under the hub to catch all the little bearings that will fall out when you lift off the gear cluster (see Figure 10-17).

Figure 10-17   Remove the gear cluster and bearings from the rear hub.

From here, you can keep on removing all the parts of the freewheel—the pawls (the "clicky" parts), the little c-ring that holds the pawls in place, and the main cap that everything mounts to. The main cap is what will be left when the gear cluster and pawls are taken off, and it will require some real force with a large pipe wrench to remove it. This cap has been forced tight by years of pedaling force, so be prepared to put some muscle into it. To remove this cap, you will be turning it in the counterclockwise direction, and it may require someone to hold the rim while you pry on the wrench handle. When you do get this cap removed, put it in your vice and grind off the ridges on each side, as shown in Figure 10-18. Don't grind it down too far or you will cut into the inside threads; just grind enough so the ridge is flush with the main body.

The reason for grinding the cap's edges is so that we can weld it directly to the hub. If you do not grind the edges, you cannot get the welding rod into the joint, since the ridges stick up past the edges of the hub. As you can see in Figure 10-19, the ground edges allow the cap to meet flush with the edge of the hub so they can be welded together. You may be wondering why this step is necessary, especially af-

Figure 10-18 Part of the freewheel mechanism is ground flat on the edges.

Figure 10-19 The freewheel cap can now be welded directly to the hub.

ter grunting on the pipe wrench to get the cap loose. Well, even though it may have been on the hub very tightly, it would eventually unscrew if you pedaled backwards with any real force once the gear cluster is welded to it, so this will prevent that from happening.

WELDING THE FREEWHEEL CAP, HUB, AND GEAR CLUSTER

Take you time and line up the welding rod properly before you close your helmet, as you do not want to slip and drag the welding rod across the spokes, or you will burn right through them. Put the other side of the hub in the vice and clamp the ground to the vice so the hub doesn't move around when you are welding the parts together. Weld a short, one-inch bead along the flat edge of the cap and the hub, as shown in Figure 10-20. The weld does not have to look nice to hold; just make sure it did fuse the two parts together rather than building

Figure 10-20 The freewheel cap is welded to the steel hub.

up on only one side. Weld both sides where you ground the edges flat, letting the hub cool in between each weld.

When you have finished welding the cap to the hub, place the gear cluster back in place and then screw the top cap back on. Don't worry about the bearings here; they will do you no good. Once the top cap is in place, weld it to the gear cluster (see Figure 10-21). Now that the cluster is welded to the end cap, nothing can ever move or come free from the hub and you now have a unicycle wheel with five or six speeds! Again, two small one-inch beads on each side will do the trick, so there is no need to weld all the way around the ring.

CUTTING THE FRAME

Now that you have made the fixed gear rear wheel, it's time for the frame. There really is not very much to make here. In fact, all you need to do is hack the down tube and head tube from an old frame, and you're done. Cut the down tube off at the bottom bracket, and the head tube at top tube, as shown in Figure 10-22. Make sure your donor frame is the correct size for your newly created rear wheel, or you will

Figure 10-21   The gear cluster is welded to the end cap and the hub.

have wasted a frame for nothing. After you are done cutting, grind off the down tube stump at the bottom bracket and clean up the end of the top tube. Guess what? You're done making the frame! Seems too easy? Just wait till you try riding it!

An old gooseneck is inserted into the end of the newly cut top tube so that the head is facing upward, as shown in Figure 10-23. Don't worry if the stem of the gooseneck does not fit snugly into the top tube, as it will be welded in place anyway. Why would a unicycle need a set of handlebars when there is no front wheel? It is simply a handle used to get started and hold the bike in place. Once the handlebars are mounted, they will be used to yank the unicycle around while you are riding, effectively steering the unit.

Try to weld the gooseneck as straight as possible to the main tube; it helps if you do this with a set of handlebars in place (see Figure 10-24).

Figure 10-22   Remove the down tube and head tube from the frame.

Just make a small tack weld at the top, and pry on the handlebars to get them straight before you finish welding all the way around. Make sure that whatever handlebars you choose fit snugly into the goose-neck so they can be held firmly in place. When you first start off, there will be a lot of pulling on the handlebars, so they need to be secure. If they are not very secure, you will have to weld them in place later.

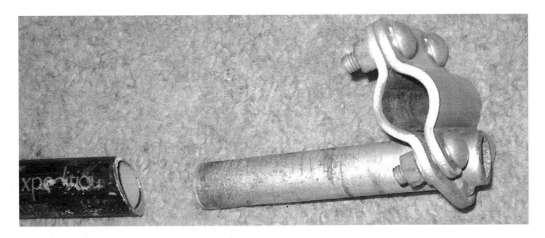

Figure 10-23   A gooseneck is inserted into the end of the top tube.

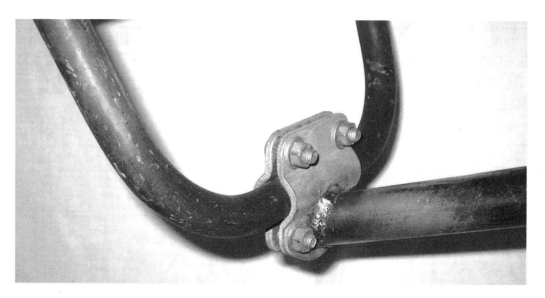

Figure 10-24 The gooseneck is welded to the main tube.

CHOOSING A CRANKSET, CHAIN, AND SEAT

You're almost done. Are you anticipating the taste of pavement yet? All you need to do now is add a crankset, chain, and seat and it's time to ride. The best type of crankset to use here is one that has the really small chain ring (common on mountain bikes; see Figure 10-25). This will allow an equal gear ratio between cranks and hub, with 1:1 being the most effective. You may end up with a front sprocket having 26 teeth and a rear sprocket with 22 but this will be close enough. Try to keep the ratio as close as possible without having the larger amount of teeth end up at the rear hub. If there are more teeth on the rear sprocket than at the front, you will pedal wildly but be moving very slowly. When you do find the best rear sprocket, make a chain to fit and put it in place.

When you are putting on the rear wheel (the only wheel), try to get the chain as tight as possible so that there is no slop as you pedal. You will be making many small corrections as you try to balance, and a loose chain will not give you fast enough feedback to avoid falling. Set the chain up and give the cranks several turns to make sure it is going to work properly (see Figure 10-26). When choosing pedals, find a set that is the same on both sides. When you first start off, you will have to swing one leg up onto the pedal very quickly, and if you try to use a pedal that has a high side and a low side, you will have a 50 percent chance of ending up with your foot on the wrong side every time. Those

Figure 10-25 Choose a crankset that includes a small chain ring.

Figure 10-26 Front and rear sprockets are chosen with a 1:1 gear ratio.

typical cheap black plastic ones that are abundant on department store bikes will work just fine.

The last piece will be the seat. A banana seat is best because it lets you start farther back on the frame so you don't have to lift yourself as high. A banana seat also lets you slide around while you attempt to keep your balance, much like a real unicycle seat. You could get away with a standard seat, but learning to ride the unit will be much harder and may require more visits from the tooth fairy.

By using a banana seat, I found that most of my falls were from the back of the bike, not nosedives, which is a good thing. Once I learn to ride the Wild Bull much better, I may put a regular seat on, just to make it more interesting and challenging to ride.

The final product is painted and ready to ride (see Figure 10-27). Notice the large aggressive tire I put on the wheel. You never know when you may want to take your one-wheeled creation out to the mountain bike trails for some off-road riding! Actually, it was the only unmatched tire lying around, so I used it, but any tire will work, even a bald one.

Figure 10-27 The completed Wild Bull, ready to toss off any rider brave enough to climb on!

Before you head out of the garage thinking that you will make it to the end of the street on your first attempt, let me give you some pointers to speed up your learning curve.

RIDING TIPS AND TRICKS

First, you need to adjust your seat so you can easily reach the pedals when sitting as far back as possible. The best place to ride is at the rear of the seat, so most of your weight will be riding over the rear wheel. You must also be able to touch the ground with one foot while the other is on one of the pedals, as this is how you will be starting off.

Look at Figure 10-28 and study the starting position. Notice the rotation of the crankset. The crank arm that will be used to start moving must be facing forward, not backward. If you try starting with a backward crank arm, you will be introduced to the pavement as though you were bungee jumping with no bungee cord! The idea is to start moving into a wheelie, not start in reverse.

Figure 10-28 Starting position is the key to taming the Wild Bull.

The arm used to hold onto the handlebars will be on the same side as the foot on the pedal, so if it's right foot on the pedal, then it's right hand on the handlebars. Huh? Holding on with only one hand? Yes, this is correct. If you try to ride with both hands, you will not be able to use your arm as a counterbalance by swinging it, and you will not ride farther than 10 feet, trust me.

Set the frame so that the top tube is about parallel with the ground, then push down on the foot already on the pedal and try to get the other foot on the pedal as the crank arm comes around. This is the hard part—starting off. If you can master getting both feet on the pedals before the unit has fallen forward, backward, or sideways, then you might just have a chance. Within two seconds from that first push on the pedals, you should be blasting down the street with one hand on the handlebars and the other waving wildly through the air just like a real cowboy on a real wild bull. Or you will just keep getting a close view of the pavement.

Never give up! I can ride a unicycle, and it still took me over two hours to get the Wild Bull under control but, eventually, it happened. You will find that "sweet spot" where your body is perfectly balanced

Figure 10-29 Taming the Wild Bull. Have patience and you will always win.

over the rear wheel, and all you need to do is swing your free arm left and right to maintain equilibrium. Notice my good "bull riding" technique, as shown in Figure 10-29? When you get good at balancing, it may even be possible to turn a full circle in the width of the street or do spins like the ones that can be done on a unicycle.

I learned a few things from building this project. First, anything you can imagine made from bicycle parts will probably work if you just go and build it. Second, anything with any combination of wheels can be ridden if you practice at it long enough.

Don't think a project will fail until you have a completed prototype. You might change your design many times but, if you keep trying, you'll be pleased with the results. The three projects outlined in the next chapter—the Kool Kat, Marauder, and Coyote—are perfect examples of my persistence philosophy.

LIFE IN THE FAST LANE

The Recumbent Revolution

The Kool Kat is a good example of the revolution that has begun in the ancient world of the standard diamond-framed bicycle—the Recumbent Revolution. The word "recumbent" refers to the reclining position of the rider on the vehicle. Rather than hunching over a little seat to reach the handlebars, you sit back in a nice comfortable seat with your legs ahead of you, as if you were sitting in your favorite relaxing chair. This recumbent position has several advantages over the upright riding position, but the main one is comfort.

I remember when I built my first recumbent-style bike and took it out for a test ride with only bare plywood as a seat. What a surprise it was when this bare wood seat turned out to be remarkably more comfortable than the gel-filled seat on my mountain bike. In fact, it was such a difference that I rode the recumbent with only the bare plywood seat for the entire year.

Another advantage of the recumbent riding position is the amount of force you can push on the pedals. Because your seat is directly behind your back, there is no limit to the amount of pressure that can be exerted on the pedals. This can be compared to using a leg-press machine at a gym. On an upright bike, you can only press down with a limited amount of pressure before you lift yourself off the seat. A physically fit person can press double his/her body weight for the short distance required for a pedal stroke with one leg, thereby doubling the amount of energy that can be delivered to the cranks. Although this type of hardcore pedaling force would only be used in racing or hill climbing, it is fun to slip past those hi-tech roadies on their multi-thousand-dollar carbon fiber bikes with your $30 home-built recumbent!

The last main advantage of the recumbent position is aerodynamics. Once you reach a certain speed, wind resistance is the major energy waster working against you while you ride. A recumbent bike puts the legs directly in front of the body and puts you in a lower position. On a windy day, even the casual rider will likely whiz past the seasoned upright rider while traveling against the wind. Pitting the recumbent against the upright bicycle will open a heated debate wherever you go as to which is the better design, and, in the end, it is really just a matter of preference. Recumbents are fun to ride, comfortable, and get attention wherever they show up, and that is enough reason for me.

Introducing the Kool Kat

The project presented here is named "Kool Kat" and both words are spelt with a "K," since I built the bike for my partner and coauthor, Kathy. Because my main ride is the "Marauder," presented later in this chapter, it was only fair that she had a performance bike to keep up to me on the long rides, as a standard upright bike could not.

The Kool Kat is not a low-racer like the Marauder and would represent the "typical" short-wheelbase recumbent, if there were such a thing. The Kool Kat has a short wheelbase because the front wheel is between the rider and crankset, compared to the Marauder's long wheelbase configuration. A short wheelbase recumbent is easier to ride and can make a much smaller turning circle than the Marauder.

For the core of this project, you will need a 20-inch frame with wheels, a few pieces of one-inch tubing for the boom, and a large crankset with a bottom bracket like the ones shown in Figure 11-1. The type of bottom bracket on the main frame is not important, as it is not used, but for the bottom bracket to be used at the end of the boom, choose a three-piece crankset over a one-piece, since these are generally of higher quality and larger diameter. You will want the largest chain ring possible for this bike, since it is capable of some real speed, and you would quickly run out of gear range with a small 40-tooth chain ring typical on small 20-inch frames. Also, the rear wheel should have a gear cluster attached, rather than a fixed-gear freewheel, or else you will not have the ability to switch gears.

Figure 11-1 A 20-inch frame and wheels are used as the base frame for the Kool Kat.

Before you cut any tubing, you need to get the proper measurement for the boom length, unless you plan to modify the design to allow a sliding bottom bracket. In my design, the two tubes that make up the main boom are a fixed length to accommodate only one rider, although it could acomodate a few different riders with only an inch or two of height difference in either direction between them. The first and most critical cut you will have to make will be the bottom boom tube. This will reflect not only the inseam of the rider, but also the placement of the cranks in comparison to the front wheel. If you get this tube welded in the correct place, the rest of the bike is easy to build.

PROPER MEASUREMENTS

Remove the original seat and gooseneck from the donor frame and add a set of 20-inch wheels onto the bike. Have someone measure the distance from the seat tube to the bottom of your foot as you sit on the frame just ahead of the seat tube (see Figure 11-2). Adding a cushion or board between your butt and the frame is a good idea for comfort's sake. The measurement should be taken while you are wearing shoes and sitting in the general recumbent position on the frame. Write this measurement down and call it "measurement A."

Now that you know the distance from the back of the seat to the farthest pedal stroke, subtract the length of the crank arm and bottom bracket shell from it. Take a measurement from the center of the hole

Figure 11-2 You will need to sit on the frame to get the bottom-bracket-to-seat distance.

that the pedal threads into to the farthest end of the bottom bracket shell. Let's call this "measurement B." Now you know the distance from the seat tube of the main frame to the end of the lower tube of the main boom. If this sounds confusing, just read through the entire chapter to see how it all goes together and it will become clear.

Once we know the distance from the seat tube to the end of the lower boom tube (measurement B), we can calculate the length of the lower boom tube, since it is welded directly to the head tube. Basically, just subtract the distance from the head tube to the seat tube from "measurement B" to get this magic number. Let's call this "measurement C." Of course, you could just do it the "crude" way and have someone hold the bottom bracket and crank in position as you sit on the frame and measure the distance from the bottom bracket shell to the head tube, but this may not be as accurate.

OK, now that all the confusing stuff is over, cut a piece of one inch or similar bicycle tubing equal to "measurement C" plus one-half inch. Why? Well, the extra half-inch is to compensate for the rounded area you will grind out of each end of the tube to make a nicer joint for welding (see Figure 11-3).

Getting the proper length of this tube is a little tricky, so it's best not to weld the bottom bracket on yet. Just give it a fairly good tack weld

Figure 11-3 Always compensate for length lost when grinding the ends of tubes like this.

in case you have to start over with a new piece of tubing. When the bottom bracket is tack welded to the lower boom tube, tack weld the lower boom tube (on the top of the joint) to the lower part of the head tube so it looks like the one shown in Figure 11-4.

The hardest part is almost done! Place a crank arm onto one side of the crank axel and turn it so it is as far away from the seat tube as possible. Measure the distance from the pedal to the seat tube; this should be the same as "measurement A," the farthest distance the rider's legs will have to reach during a pedal stroke.

If this distance seems to agree with your measurements, then you only have to worry about one more thing—the clearance between the pedals and the front wheel. Go back to Figure 11-4 and look at how the crank arm is positioned as if it were passing the front wheel. The last thing you want is to have the crank arm or pedal hit the front wheel during a turn, as it would lock up the steering and cause you to crash. If the crank arm is hitting the wheel, then pull up on the tack welded tube until there is at least two inches of clearance between the closest part of the arm to the front tire.

When both the distance of the bottom bracket from the seat tube and the clearance from the crank arm to the tire are set correctly, go

Figure 11-4   Lower boom tube and bottom bracket tack welded in place.

ahead and weld the lower tube in place and then weld the bottom bracket to it, paying attention to alignment. You can now go take a well-deserved break because the rest of the building process will be a breeze!

THE UPPER BOOM TUBE

Adding the upper boom tube is just as easy as cutting a tube that will fit, as shown in Figure 11-5. As you will see, the piece that meets the top of the bottom bracket shell needs a lot of "meat" ground out to make a good seem, so cut a little extra tubing to work with. This basically forms a triangle (the strongest possible structure) between the head tube and bottom bracket. While you are welding these tubes in place, make sure that everything lines up straight, especially the bottom bracket, or you will have chain management problems. Take your time, and eye up your work from all angles.

Once the main boom is completed, you are almost done with the frame. You can now cut off the remaining seat tube since we will be replacing it with a longer, more reclined tube. Your frame should look like the one in Figure 11-6 if all went well.

Figure 11-5 The upper boom tube creates a strong main boom.

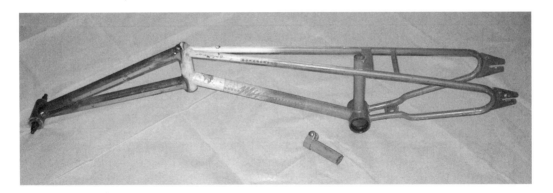

Figure 11-6 Completed main tube welded to the frame.

ADDING THE CHAIN

Now put the crankset, front forks, and wheels on the bike and add a rear derailleur. We will now add a chain to the Kool Kat. You will probably need two or more chains joined together to make one long enough to reach from the crankset to the rear wheel. Don't try to may the chain too tight at first because we will be adding an idler pulley to pick up the slack and keep the chain out of the path of the turning wheel.

Referring Figure 11-7, you will notice that the return chain's path will interfere with the front wheel if it were turned. There are actually some manufactured recumbent bikes that just let the chain rub on the wheel, but these are mainly used on racing circuits rather than on the street. If you left the chain like this and had to make a fast, sharp turn you would either lock up the front wheel or knock the chain off.

The return chain will be picked up out of the path of the front wheel by an idler pulley, as shown in Figure 11-8. This is a small wheel similar to a V-belt pulley. The return chain has no force applied to it while pedaling, so any idler will work fine, even a plastic one. You can find idler pulleys in some hardware stores, or any store that sells gym equipment (they are used to manage cables).

If you are adventurous, you could make an idler pulley from an old inline skate wheel by placing it on an axle held in a vice and then grinding a slot wide enough for the chain out of the middle of the wheel. Whatever idler you intend to use, make sure it can spin freely without resistance or annoying squeaks. The best type of idler to use is one with bearings already in place; you then only need to weld a bolt to the frame to hold it in place.

Find a place on your frame for the idler pulley that will not only keep the return chain from hitting the front wheel, but also stop the

Figure 11-7 The return chain needs to be moved up to avoid hitting the front wheel.

Figure 11-8 An idler pulley will keep the return chain out of the path of the front wheel.

chain from hitting the top of the fork legs. The best place would be along the lower boom tube near the head tube. The drive side of the chain does not need any management as it makes a straight line directly from the front chain ring to the back wheel (the most efficient path). When you find the optimal place to position your idler pulley, weld the appropriate-size holding bolt to the frame and connect the pulley, as shown in Figure 11-9.

When the idler is in place, hold up the rear of the bike and turn the cranks to make sure there is no chain rubbing. If the return chain is rubbing or banging on the frame near the rear of the bike, a simple guide can be made from an inch or two of garden hose bolted to the frame. Remember, this will only work on the return chain, since it caries no real tension, if you try this on the drive side of the chain, it will surely fail.

MOUNTING THE SEAT

Now you will need to add a short tube where the original seat tube used to be to mount the back of the new seat. The angle of this tube

Figure 11-9 A bolt is welded to the frame to carry the idler pulley.

will determine how laid back you will sit on the bike. An angle of 20 degrees or so is about right, but feel free to recline the seat to any angle you feel most comfortable with.

You can find the best angle by sitting on the bike at whatever angle you feel most comfortable with, and have someone mark this down for you with an angle finder. Keep in mind that your knees and the handlebars will be in front of you, and if you sit back too far, they will block your view of the road ahead. As shown in Figure 11-10, this short tube is welded directly over the cut of seat tube of the original frame. A length between 10 and 12 inches will work just fine.

Once the seat tube extension is welded in place at the appropriate angle, you should add some extra support from the top of the tube to the back of the frame. Since you can deliver extreme amounts of force to the pedals in the recumbent position, the seat back must be strong enough to handle this, and without some extra support it could fail. As you will see in Figure 11-11, I used the seat stays cut from an old

Figure 11-10 A short tube is added to support the back of the new seat.

Figure 11-11   The rear of the seat must be supported to withstand extreme pedaling forces.

frame, but any small tube would work fine. Try to place this part so that it forms a large triangle between the top of the seat tube extension and the rear of the frame. Once in place, no amount of pedaling forces will ever be able to bend the seat tube.

If you are using a cut set of seat stays as I did, make sure the small piece of tubing that holds the two legs together is not in the way of the tire when it is mounted on the bike. If this happens, you can remove this small tube after the two legs are welded in place, as it has no real purpose on the bike besides maybe being a good place to bolt on a fender.

Now you are ready to make a seat for your Kool Kat. You will need a few plates to bolt the wooden seat down (see Figure 11-12). Two pieces of 1/8-inch-thick plate with holes drilled on each side will do the trick. Make the plates about 6 inches wide so there is at least 4 inches of metal on each side of the frame. Drill small holes at each end of the plate so that you can use wood screws to hold the wooden seat to the

Figure 11-12 Thin steel plates are used to hold the wooden seat in place.

plates. One of my plates is wider because of the dual tubes on the base of the frame where the seat bottom will be mounted.

Since the two pieces of plywood used for the seat are glued or joined together with metal clips, you only need to support the seat to the frame at the top of the seat tube extension and in the lower part of the seat. Position your two seat mounting plates as shown in Figure 11-13. Make sure the plates are not slanted or your seat will tilt to one side of the bike.

ADDING THE GOOSENECK

The long gooseneck needed for this project is made by adding a piece of one-inch electrical conduit in between a cut-in-half gooseneck. The goal is to position the handlebars just above the highest level of your knees while you are pedaling. In my design, a piece of tube measuring 14 inches was just right, but you may need more or less, depending not only on the donor frame geometry, but the size of the rider. Once complete, there will be about 3 inches to work with by adjusting the height of the gooseneck within the head tube. As shown in Figure 11-14, I used a chopped BMX gooseneck for the bottom and a standard mountain bike type for the top. This design made it easy to weld the extension tube directly to the base of the gooseneck and utilize the original bolt and wedge.

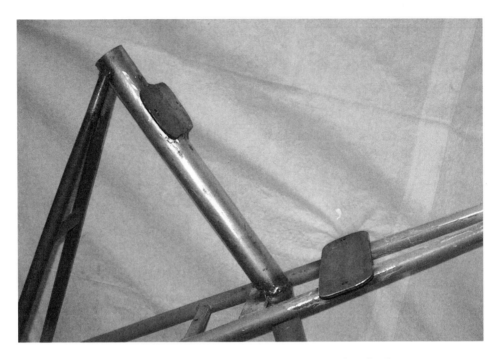

Figure 11-13 The two seat plates are mounted to the frame.

Figure 11-14 A gooseneck is extended with a piece of one-inch electrical conduit.

The easiest way to find the proper height for your handlebars is to sit on the bike and pedal while someone measures to the top of your knees. Make sure that you sit on a pillow or some padding rather than just the bare plywood when you take this measurement or it will be off by a few inches.

The final gooseneck should be straight and strong. Grind all welds and check them over to make sure there are no cracks or flaws. You don't want to have your steering fail as you blast into a sharp 40-mile-per-hour corner! A set of handlebars with a downward curve (see Figure 11-15) feels most comfortable on this type of recumbent bicycle. You may also want to cut a few inches off each end to shorten the width of the handlebars as well, but make sure you leave enough room for the grips, brake levers, and shifters.

ADDING ACCESSORIES AND THE SEAT

Now that you have all the welding completed, you can paint the frame and accessories (see Figure 11-16). Because the frame is so small, a single can of spray paint is more than enough to give two good coats. You will also want to paint the extended gooseneck so the welded areas do not rust.

While you wait for the paint to dry, find some half-inch plywood and cut out two pieces to make your seat base. A width of 8 to 10 inches will work nicely, and a length of 10 to 12 inches for the base should also work. The rear of the seat is made to be as tall as the top of the extended seat tube to look good. You can make your seat any size you like, but don't go too long on the bottom part, or you will have a tough time putting your feet down to get off the bike when you stop. As you will see in

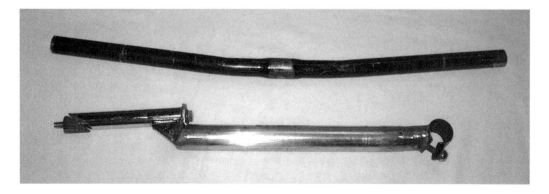

Figure 11-15 Choose handlebars with a slight downward curve for best results.

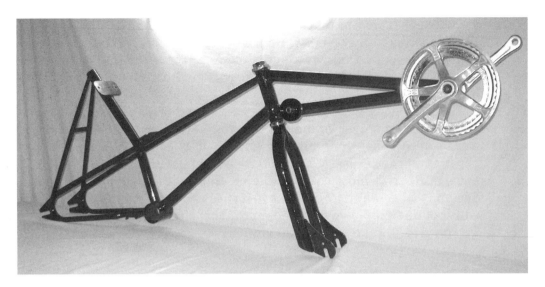

Figure 11-16   The Kool Kat is painted and looking "kool."

Figure 11-17, I also tapered the ends of the seat so it would not look so blocky.

As for padding, any stiff foam will work. I used a 2-inch thick type of hard foam that is used to make wheelchair seats. You can get small pieces of foam at most fabric stores in small amounts, or by cannibalizing an old couch cushion. If you end up using soft foam, you will have to cover the seat with some type of vinyl or cloth to keep the foam from ripping with use. Try to avoid using seat coverings that

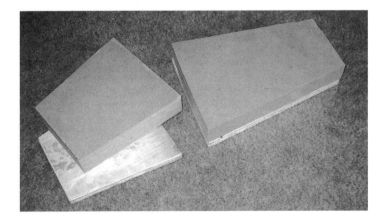

Figure 11-17   Foam is added to the plywood seat base.

don't "breathe" or you will find yourself soaked with sweat after a long hard ride.

When your paint has dried, grease all the bearings and reassemble the bike. The two seat bases are then screwed down to the seat clamps with heavy wood screws. I also used bent shelf clamps screwed to each side of the seat bases where they meet up with each other just to make sure everything would be very secure. A gear shifter and brake have also been added to the finished and painted unit (see Figure 11-18).

RIDING TIPS AND TRICKS

Riding the Kool Kat takes a little getting used to if you have never owned a short-wheelbase recumbent before but, overall, it is a breeze to learn. The strangest part about riding this bike is how the bottom bracket stays in one place as you turn the handlebars. When you first start out, you will tend to over-steer as your eyes play tricks on you, but once you get used to the stationary bottom bracket, you will be a lot smoother starting off.

To start off on the Kool Kat, set the pedal on your "favored" side to the topmost position. Your favored side is the foot you will put on the

Figure 11-18   The Kool Kat is ready to prowl the streets.

pedal to start the bike in motion as the other foot remains on the ground.

Sit balanced in the seat, take the handlebars, and with your favored foot on the pedal, push off to get the bike rolling then bring the other foot onto the pedal. To make a smooth and successful start, you must trust yourself and the bike. If you think you are going to lose your balance, you will tend to start off too slowly and end up making your wish come true.

Once you manage to get going, the rest is smooth sailing. The bike will ride much like a regular bike besides your very relaxed seating position. Once you go around the block a few times, you will notice a few advantages a bike like this has over the "old fashioned" upright bicycle. For starters, you can really get moving fast from a standstill, and there never seem to be enough gears once your legs get used to the new riding position.

Keeping up to traffic in the city limits is not hard at all when you feel at ease on the Kat. Also, you can keep on pouring the juice into the pedals as you go around corners because the pedals are so high off the ground they could never scrape as they would on a regular bike.

Once you build a recumbent like this, you may never want to ride a regular upright bicycle again! Remember that sore butt you used to get

Figure 11-19 Kathy puts the fast and comfortable Kool Kat to the test.

when you first got back to cycling after the winter? Say goodbye to that soreness and hello to the Kool Kat. The only disadvantage I have found with a recumbent bike so far is the amount of waiting you have to do as your riding buddies try to catch up to you on their "wedgie" seat bicycles!

The next project, the Coyote, is another recumbent that's built for comfort, but can also achieve high speeds.

The Coyote

COYOTE CHARACTERISTICS

The Coyote is two-wheeled savage beast bred for only one thing all-out speed. If you want a nice tame bicycle for those calm leisurely rides, then this isn't the one for you! Because of the Coyote's extremely low seat height and high bottom bracket, the rider is able to really exert some force into the pedals. Also, the aerodynamic position of your body combined with the ultralight frame and components make this one speedy machine.

The Coyote is based around a single tube frame with two ultra-light-weight 27-inch (70 cm) rims at each end. Because the unit is front-wheel drive, the chain does not have to be routed under the seat, so the position of the rider can be very close to the ground. This is a very minimal racing bike with only seven gears, a single chain ring on the crankset, and one rear brake. Although the Coyote is fairly well be-haved at high speed on a smooth surface, I would not want to take it out into traffic or on a busy street. This vehicle is made for track or indoor racing. If zipping along only two inches off the ground at ridiculous speeds sounds like a good time to you, then let's begin!

IT'S ALL ABOUT THE PARTS

First, you will need a good set of 27-inch wheels—one front and one rear. These are not the kind of rims you find on a typical mountain bike, which are a lot heavier and are marked "26 inch" on the rim. The ones you want are made for those lightweight road/speed bikes and were very popular in the 1980s on those "ten speeds." Try to find a good-quality aluminum set, as we want to keep the weight of the Coyote down to a minimum. Also, the higher the tire pressure, the better the rolling resistance, so try to find a tire that can take between 80 and 100 psi.

When you have found a suitable set of wheels, you will need a pair of front forks and a pair of chain stays that will be able to take the 27-inch wheels. A pair of front forks from a 26-inch mountain bike will work just fine. As shown in Figure 11-20, I used front forks from a mountain bike and cut the chain stays from the same bike's frame. Cut the chain stays out so that the bottom bracket and rear dropouts are still attached.

Choose a bottom bracket designed for a three-piece crankset rather than a one piece. The three-piece crankset is usually lighter, has a larger chain ring (50 or more teeth), and is made of lighter materials. Because the Coyote has only a derailleur at the front wheel, a single chain ring is all that is needed. The larger the main chain ring, the faster you will go.

CUSTOMIZING THE FRONT FORKS

Cut the ends of the front fork dropouts into a diamond shape, as shown in Figure 11-21. They will be welded to the donor chain stays at the

Figure 11-20   A pair of 27-inch wheels and tires, front forks, and chain stays.

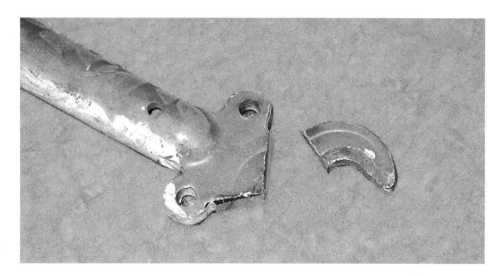

Figure 11-21 The tips of the forks are cut into a diamond shape at the ends.

joint where the seat stays used to be, and this diamond shape will make it easier to adjust them once they are tack welded into place.

The front forks will be carrying a rear wheel, so they will need to be stretched apart enough so that the newly cut dropouts mate with the chain stay dropouts. Stand on one leg of the front forks and carefully pull up on the other leg, forcing the legs apart a little at a time. Repeat this until the fork dropouts are the same distance apart as the chain stay dropouts (rear dropouts). Once you have the front forks widened to the same width as the chain stays, tack weld the tip of the front forks to the top of the rear dropouts where the seat stays used to be. There should be about 14 inches from the bottom bracket to the lower part of the fork stem, or crown. Make sure the joints at the fork tips and bottom bracket are straight and in the same place at both sides. Your tack-welded unit should look like the one in Figure 11-22.

Before you go any further, put in the rear wheel to make sure everything is going to fit. The wheel should be in the center of both the front forks and the chain stays at the same time. If something seems out of alignment, break the weld on one side of the forks and redo it until the wheel sits perfectly straight. This bicycle is going to be traveling at some fairly high speeds, so you don't want to have a badly aligned wheel.

When the two parts are joined properly, cut a 14-inch section of tube from any part of a donor frame. A one-inch diameter piece of tube or

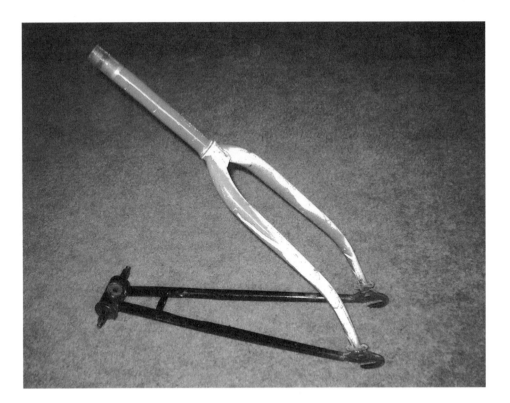

Figure 11-22 The forks are widened, then welded to the top of the dropouts.

conduit will work just fine. This tube will be welded between the bottom bracket and the fork crown (see Figure 11-23). The tube should be welded to the heavier area of the forks where the two legs meet, just below the bearing ring. The bearing ring should be removed before welding just in case you slip. Tack weld the tube at both ends, then put the rear wheel back in place.

As shown in Figure 11-24, the rear wheel should have about an inch of clearance between this new tube and the fully inflated tire. If everything is going good, you can remove the front wheel and finalize the welds at the fork tips and both ends of the 14-inch tube. Remember to check your wheel alignment. You don't want to crash into the wall during a race!

You will also notice that the dropouts will be upside down in the final design. This has not been a problem as long as the wheel nuts are on tight. If this setup worries you, cutting the dropouts off and rewelding them on the other way is an option, as well as using those special washers that have a hook to stop the wheel from moving.

Figure 11-23 A short tube joins the forks to the bottom bracket.

THE HEAD TUBE AND DOWN TUBE

When your front end is completed, you will need to find a head tube that will fit the front forks properly. Cut it from the donor frame and grind off all the excess metal, as shown in Figure 11-25, until it is perfectly bare. Put the head tube, bearings, and all other hardware onto the front end as they will be in the final design. Make sure the bearings match the cups correctly or the bike will have stiff handling.

The next tube to be cut will be the front part of the main frame (I will call it the down tube from now on). A two-inch-diameter, thin-walled electrical conduit is what I used, but any round tube with a diameter of no less than two inches and a wall thickness of at least 3/32″ will work. Muffler pipe or square tubing of this size can also be used, but conduit is easier to find and will cost less.

Cut this tube (down tube) to a length of 24 inches, then groove out one end so it mates with the head tube (see Figure 11-26). The goal

Figure 11-24 The wheel should have adequate clearance between the tire and frame.

Figure 11-25 The head tube should be the correct size for the fork stem.

Figure 11-26 The down tube should be no further than one inch from the front wheel.

here is to set it up so that the closest part of the front tire is about one inch from the down tube. This will put the front forks at the correct angle when the frame is complete. The tube should also be placed so that it is closest to the top of the head tube. This will create a steeper angle, allowing the seat to be lower to the ground.

When you have the down tube ground out to fit on the head tube with an inch of clearance between the front tire, weld it in place. Make sure the down tube and head tube are perfectly aligned with each other, or your front forks will be on an angle from the rest of the frame. It is best to leave the front forks connected when making this weld so you can look between the fork legs to see if the down tube is straight as you weld it all the way around.

Once the head tube is welded to the down tube, you will need to cut the other tubes that make up the rest of the frame. I cannot give you a measurement here, as this is a personalized vehicle made to fit your leg length without having to make any seat adjustments. The tube that determines the size of the rider is the lower horizontal tube that runs parallel to the ground (see Figure 11-27).

Figure 11-27   Laying out the frame tubes in order to figure out the proper length.

DETERMINING THE PROPER LENGTH

When you sit in the bike, your back should be firmly pressed against the rear of the frame (back tube) while you push on the cranks. On your farthest pedal stroke, your knee should be almost locked. If you frame is too long, you will have a hard time reaching the pedals, and if it is too short, it will be very hard to pedal, if not impossible.

The way I figured out the proper length of the frame was by sitting in a chair and stretching out one leg along the ground as far as it could go. The length from the underside of your outstretched foot (wearing shoes) to the farthest point on the chair's bottom seat should be the same distance as the farthest pedal to the joint between the lower frame tube and back tube.

I recommend that you lay out the bike on a flat surface with a straight line or stick representing the ground, as shown in Figure 11-27. Then, you can set the angle of the front end (about 35 degrees) and lower frame tube. If you make the lower frame tube longer than you will need, weld it in place. Then you can just tack weld the rear seat back tube in different areas until you find the position that fits you best. While doing this, remember that the seat and padding will also add about one inch on the bottom and back of the seat, so include this fact into your calculations. When you do find the "magic" length to make your frame, the rear frame tube (the one your back is against) will be welded in place at about the same angle as the front end. I used 35 degrees, but since this is an experimental vehicle, feel free to try out different angles. If you go too far back, you will only see your knees while you ride, and if you don't go back far enough you will have your knees in your face as you pedal.

Figure 11-28 shows an alternative design for the back of the frame using a piece of curved tube cut from a factory-bent conduit elbow. This design looks a little better due to the curve in the frame, but other than that, it works out the same as the straight-tube design shown in Figure 11-27. Again, the length of the rear frame tube is not critical and, in my design, the tube extends 18 inches up from the joint at the bottom tube. Anything from 14 to 24 inches would work just fine, so feel free to experiment.

THE REAR TRIANGLE

The rear triangle is made from a pair of forks and the two halves of a seat stay (which you should have left over from your donor frame). The

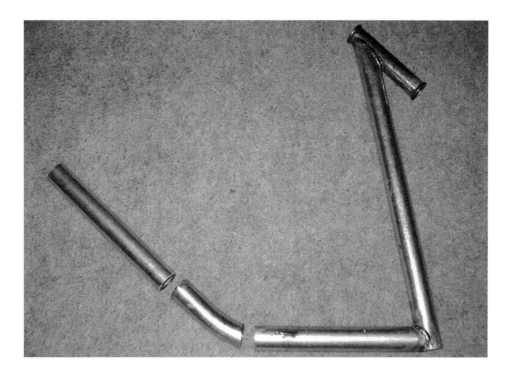

Figure 11-28 An alternate rear frame design using a factory-bent elbow.

forks must fit the 27-inch wheel just like the ones on the front of your bike, and can be of any style. The end of the fork stem is cut at an angle (see Figure 11-29) so that it can mate with the rear of the frame. The goal is to keep the wheel as close to the rear frame tube as possible while maintaining the lowest possible frame. If you refer back to Figure 11-27 or ahead to Figure 11-37, you will see the basic position of the frame and wheels.

Start by welding the end of the fork neck to the area where the bottom frame tube meets the rear frame tube (see Figure 11-30). Remember to only tack weld the joint at first so you can get the rear wheel aligned with the frame. This task is easier to do with the wheel in place. Look down the end of the frame to make sure the wheel is not leaning to one side or the other and make whatever adjustments needed before you add the final weld. Try to get this the forks welded in as straight as possible or you will have to constantly steer just to go in a straight line.

Once you have made the weld at the fork stem and frame, you can weld each leg of the seat stay from the fork tips to the top of the rear

Figure 11-29   A set of front forks and two seat stay legs make up the rear triangle.

Figure 11-30   The fork stem is welded to the underside of the frame.

frame tube (see Figure 11-31). This is an easy task because the forks should already be perfectly aligned. As one last check, put in the rear wheel to make sure the tire does not rub on any part of the rear of the frame before you do any welding. The stays are welded to the top of the dropout area on the fork tips and to the top of the rear frame tube. Weld them to the outside of each side of the rear tube so there will be adequate width between them for the rear tire. After you are done, place the rear wheel back into the frame and check alignment one last time.

THE SEAT AND MOUNTING PLATES

Now you will need eight small square plates like the ones in Figure 11-32, with a hole drilled in the center to mount your seat to the frame. 1/16-inch thick plate at about 1.5 inches square will do the trick. You can also cut 90-degree shelf brackets in half and use them as well. Using wider plates with holes at each end will also do the job, but this will put the seat up a little higher on the frame.

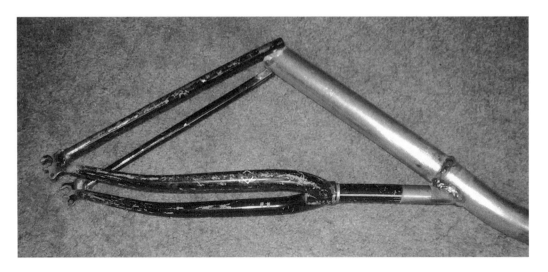

Figure 11-31   The two seat stay halves complete the rear frame triangle.

The plates are mounted so that each half of the seat will have four wood screws holding it down to the frame, as shown in Figure 11-33. I placed a plate on either side of the round tube so that seat boards would be resting on both the mounting plates and the top of the tube. There is nothing special about welding the seat plates to the frame besides making sure they are not leaning from to one side or you will fall out of your seat.

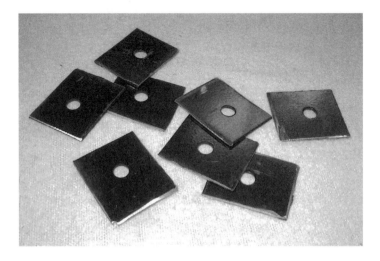

Figure 11-32   The seat mounting plates are made from thin 1.5 inch square pieces.

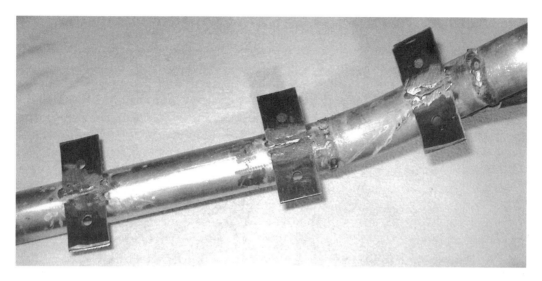

Figure 11-33 The seat-mounting plates are welded in pairs to the frame tubes.

The actual seat is made from two pieces of 10-inch-wide, half-inch-thick plywood with some type of hard foam glued to the top. The foam does not have to be very thick as it is only there to take the edge off the hard seat as you are racing around. As shown in Figure 11-34, the ends of the seat boards are cut or rounded at the corners, then screwed down to the mounting plates with large wood screws.

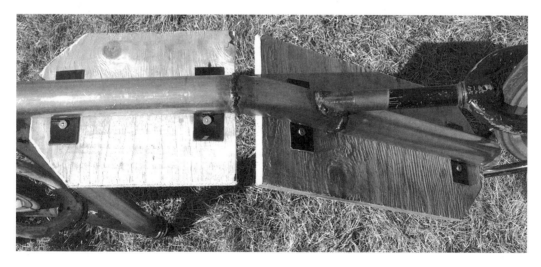

Figure 11-34 The plywood seat is securely mounted to the frame with large wood screws.

ADDING THE DERAILLEUR, BRAKES, AND HANDLEBARS

You're just about ready to race! All that is needed is a derailleur, brakes, and handlebars and you will be passing upright bikes like they were standing still. As you will see in Figure 11-35, the derailleur is positioned so the body is in front of the bike, since the side of the chain that actually does the work is behind the forks, closer to the rear of the bike. At first glance, this may look strange, but it actually makes total sense. Just imagine the photo rotated 90 degrees clockwise, and you will see that it will then look exactly like the rear end of a regular bike.

When your derailleur is in place, make sure you adjust the two limiting screws so that the full range of gears can be reached without the chain overshooting the smallest or largest gear. One thing you don't want is for the chain to the derailleur to get caught in the spokes and lock up the front wheel at high speed. Ouch!

A rear brake provides plenty of stopping power, and can be mounted right on the rear forks where it was on the original bike. Since the

Figure 11-35   The derailleur is mounted so the body faces the front of the bike.

forks were used without any modifications to the length or dropouts, the brake will work just as it did on the original bike. As shown in Figure 11-36, the brake is bolted through the hole in the fork crown and mounted so it is on the topside of the forks. The brake pads should be adjusted so they are flat against both edges of the rim when fully closed. Do not allow the brake pads to rub on the tire, or you could blow the tire out by burning a hole through the sidewall.

As for handlebars, there are two ways to go. You can use a short, straight set cut to be as narrow as possible like I did (see Figure 11-37), or you can use a large, curved set that comes down closer to each side of your body. The short, straight-style handlebars will give you a more aerodynamic position since your arms will be closer to your body, but the bike will be harder to ride due to the countersteering forces induced by hard pedaling. I found this a bit awkward at first, but after a few rides forgot all about it.

The larger-style handlebars, such as ones found on "beach cruiser" style bikes, will allow you to have your arms in a more relaxed position and give you a little more control over the bike. If you are just learning to ride a low racer like this, then this may help you get used to the strange feel of such a machine.

Figure 11-36   The rear brake is mounted on the forks like it was on the original bike.

Figure 11-37 The completed Coyote low racer ready to shred some pavement.

Mount the brake and shifter lever in a place that allows proper operation without needing to take one hand off the handlebars (something that would end badly).

RIDING TIPS AND TRICKS

Before you ride, let me tell you what to expect. The first time you attempt to get moving you will probably wander all over the place then fall out of the seat due to the strange feel of pedal-induced steering. As you push on the pedals, the steering wants to move a little in the direction of force from your legs. To overcome this, you have to learn to instinctively counteract this force by pulling on the opposite side of the handlebars from the leg that is pushing. This may sound hard to do, but after about 15 minutes, I was blasting around the street at high speeds and able to make nice, sharp figure eights with very little effort.

As you will soon realize, the Coyote can reach some really scary speeds and fly around corners without slowing down. The unit is very aerodynamic, light, and puts you in a good position to really push on the cranks. The downside is that it is not a good machine for visibility in traffic, or comfortable for long rides, so it is really a track bike made for the racing circuit. I have taken the Coyote out for street rides, but

Figure 11-38   Learning to ride the Coyote is not hard, but will require some practice.

do not feel safe in busy traffic. I do have to admit that it is fun creeping up on Spandex-clad upright wedgie riders, then leaving them three blocks behind as I put all my energy into the cranks! When you first take this bike out for a few rides, you will be sore in places you didn't even think you had muscles.

Well, how do you like the feel of speed? It's amazing how fast you can go on a human powered vehicle that weighs less than a single rim from an automobile, isn't it? With only a light, streamlined shell around a bike like this, it is possible to reach speeds of over 68 miles (110 kilometers) per hour! Think I'm kidding? Look up the name "Sam Whittingham" in a search engine and you will see that it has been done; and this is from a standing start on a flat road without any aid from a motor, hills, or wind. It's truly amazing!

The Marauder

Although it may only look like a pole with a wheel at each end, the Marauder represents the most engineering I have ever done on a bike project and it is not only a great machine for comfort and handling, but it is by far the fastest moving object in this book. The Marauder is the result of six months of redesigned frames and originally began life as a tricycle with two 20-inch front wheels and a rear 20-inch wheel. After four or five more major frame and steering alterations, it has become the sleek and fast low racer presented here.

My main objective while designing the Marauder was to create a low racer that had all of the positive characteristics that make a bike like this fast and fun, but none of the negative characteristics that would make it unpractical for street driving. I'm not a big fan of driving around in circles on a small racing track, but do enjoy going as fast as I can with only my own power. So, I needed to build a bike that would be able to provide both speed and proper handling on the street.

Many commercially available low racers on the street have a problem with the driveline. Because they are mainly designed for racing on smooth bicycle tracks and indoor velodromes, there is no real need to be able to make sharp turns or have a large field of view. As long as you can see the track and whom you are about to pass, that is usually sufficient. Therefore, most commercially available low racers have a short-wheelbase design that this places the front wheel in between your seat and the pedals. The front chain ring is placed ahead of the front wheel, so the chain must pass alongside the front wheel to get to the gear cluster on the rear wheel. Also, unless you have ultralong legs, there will be some crank arm overlap with the front wheel as you pedal. Having a chain running only an inch past the front wheel and a crank arm that can stop you from steering may seem a little dangerous, but since you only making the same small turns on a racing circuit it's not a serious problem.

The Marauder is designed to avoid the visibility and steering problems associated with short-wheelbase low racers by placing the front wheel ahead of the cranks, and placing the bottom bracket low enough so that you can see more than your feet in front of you. The disadvantage of a longer bike is that the turning radius will be larger, but you can still do a U-turn on a narrow city street, if necessary. I would certainly choose a larger turning radius over having my crank arm jammed into the front wheel as I was trying to avoid hitting something or someone.

MARAUDER CHARACTERISTICS

Most of the frame for the Marauder is made from 3/32-inch-thick, 1.5-inch-diameter square or round tubing. Square tubing will be easier to work with, especially when welding the angles, but I have seen this frame made with round tubing and it turned out just as good as my square-tube version. Besides the square tubing for the main frame, you will also need two head tubes, a three-piece crankset bottom bracket, chain stays cut from a 26-inch frame, and two pairs of forks, one for a 20-inch wheel, and the other for a 26-inch wheel.

Refer to Figure 11-39 to familiarize yourself with the basic frame parts and dimensions. Don't become overwhelmed if it seems like there is a lot more information to contend with compared to the other projects in this book; the letters are mainly reference points to make it easier to refer to certain areas of the frame. The Marauder shares almost none of the familiar geometry of a regular double-diamond frame, so I will call many of the frame parts by the letters assigned in Figure 11-39.

Although you may want to change some of the measurements and angles to suit your own building style, many of them can be used exactly as shown to build your own Marauder with the exception of tube "G," as this determines the proper length of the frame for a rider.

For reference, here are the lengths, angles, and names for each letter of Figure 11-39:

A   Rear forks (made from a 26-inch set of front forks)
B   Chain stays (cut from a standard 26- or 27-inch frame)
C   Seat back tube (15 inches of 1.5-inch square or round tubing)
D   130 degree angle between tube "C" and tube "G"

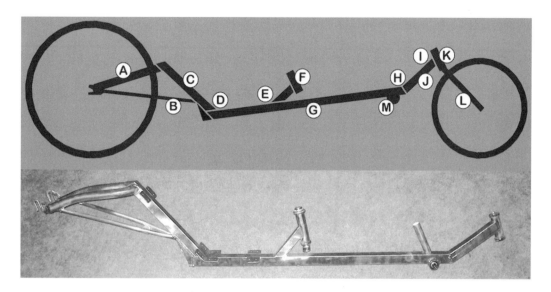

Figure 11-39 Each part of the Marauder's frame will be assigned a letter for reference.

E 145 degree angle between the remote steer tube support and tube "G"
F Steer tube (standard head tube cut from a bicycle frame)
G Main tube (1.5-inch, 3/32-thick square or round tubing)
H 145 degree angle between tube "G" and tube "J"
I 105 degree angle between tube "J" and head tube "K"
J Front tube (9 inches of 1.5-inch square or round tubing)
K Head tube (standard head tube cut from a bicycle frame)
L Front forks (standard 20-inch bicycle forks)
M Bottom bracket (standard three-piece bottom bracket)

The first thing you need to do is determine the correct length for the mainframe tube "G." This length determines the proper position of the bottom bracket, depending on the height or inseam of the rider. Table 11-1 provides a fairly accurate comparison between rider height and the length of the main tube, although you may want to find this measurement out for yourself.

CUTTING THE TUBES

When you have determined the correct length of the main tube "G," cut it off square at both ends. Now you can cut the 15-inch length of tubing for the seat back tube "C." As shown in Figure 11-40, tube "C" rests on top of tube "G" to form an angle "D" of 130 degrees, so it will need to have the end cut at the proper angle. Weld tube "C" so that the bottom

Table 11-1 Rider height and main tube length

Height of rider	Main tube length
5' 4"	34"
5' 5"	35"
5' 6"	36"
5' 7"	37"
5' 8"	38"
5' 9"	39"
5' 10"	40"
5' 11"	41"
5' 12"	42"
6' 0"	43"
6' 1"	44"
6' 2"	45"
6' 3"	46"
6' 4"	47"

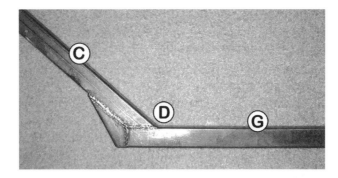

Figure 11-40   The main tube "G" and seat back tube "C" are joined and the gusset is added.

rear corner meets the top rear corner of tube "G" (see Figure 11-40). The gusset is added for extra strength and is made from a piece of tubing, the same as the tubing used for the frame. The gusset is necessary here because this is the joint that will support most of the rider's weight.

Once you have tubes "G" and "C" joined, cut the 9-inch length of tubing needed for the front tube "J." This tube will join the head tube to the frame and help lower the bottom bracket. It would be possible to eliminate this tube and just extend the length of the main tube "G" right to the head tube, but this would place the bottom bracket quite a bit higher, and you would have to look over your toes as you ride. As shown in Figure 11-41, tubes "G" and "J" are joined so that an angle "H" of 140 degrees is formed. Grind the ends of the tubes to the proper angles for joining, and then weld them together.

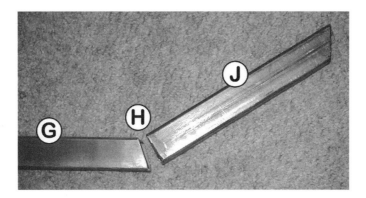

Figure 11-41   The main tube "G" and front tube "J" are joined.

The remaining end of front tube "J" is now grooved out to take the head tube "K," as shown in Figure 11-42. An angle of 105 degrees is formed at the joint labeled "I" (see Figure 11-42). This angle determines the steering angle of the forks in the final design. If you want to experiment with this angle, then only add a small weld on each side of the joint for now, so you can grind it off later for alteration. An angle of 105 degrees from the front tube was chosen because it puts the front forks at about the same angle that they would be on a regular bicycle, and this has proven to be a good choice for handling and control on the Marauder. As you reduce the angle of the forks so that they become closer to horizontal, you create steering with a much faster response, and this can give you a "twitchy" feeling when you ride, which is scary at higher speeds.

To create the rear of the frame, a set of 26-inch mountain bike or similar heavy forks are cut so both legs are removed from the stem, as shown in Figure 11-43. Using a thin cutting disc for your grinder, cut the legs as close to the joint as possible to save as much material as you can. The cut ends of each fork leg should be straight and of equal length so when they are welded to each side of the frame tube, they will be of equal distance from the center. Also, try not to damage the stem or threads from the forks as you cut because they will be used to create part of the front steering system.

Try to find a heavy set of forks to use in this part of the frame. Each leg will be supporting a lot of weight. The thin tapered forks you find on light-duty mountain bikes and road bikes may not be strong enough, especially at the tips where the rear wheel mounts to the dropouts. The best type of forks to choose are the ones with very little taper near the tip, looking more like a single length of bent tube, like the ones shown in Figure 11-43.

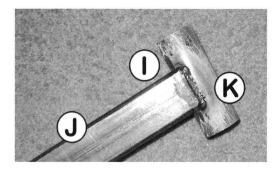

Figure 11-42   The head tube "K" is joined to the front tube "J."

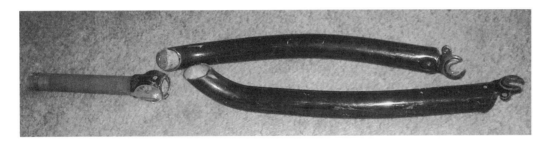

Figure 11-43 Cut the fork legs from the stem as close to the joint as possible.

JOINING THE FORKS

Get ready for the hardest part of the entire building process—joining the fork legs to the frame. This can be a difficult task because the fork legs have to be positioned so that both are at the same angle. Both must have an equal distance from the frame center, and the lowest part of the frame should not have less than five inches of clearance to the ground when the rear wheel is in place. This may seem like a daunting task to complete without a complicated frame jig, but it's not; it's only a test of patience. I have successfully made three of these frames now, so I have found an easy way to get the job done.

First, place the cut for the legs into the rear wheel you intend to use and tighten up the nuts so both fork legs are held securely in place at the same angle to each other. Next, place the front wheel and forks into the head tube and lay the frame on the ground, as shown in Figure 11-44. A long stick or string is placed between the wheels to represent a ground reference point to gauge how much clearance there will

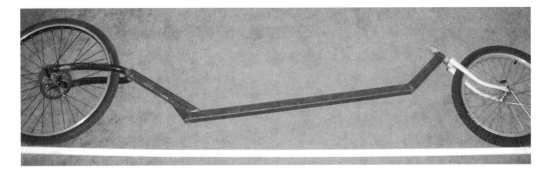

Figure 11-44 Use a ground reference point when jointing the fork legs to the rear of the frame.

be between the lowest point on the frame and the ground (5 to 6 inches is a good amount).

With the frame set for proper clearance, place the top of the fork legs into position at the top of the seat back tube "C" (see Figure 11-44). Don't worry about the side-to-side alignment right now as this will be adjusted later—just be concerned with the height of the frame and position of the fork legs.

Once the fork legs are placed in the proper position, add a sturdy tack weld to the top leg of the fork (the one facing you). With this sturdy tack weld holding one of the legs to the frame, you should be able to move the frame onto a workbench to clamp it into a vice without disturbing the angle of the fork leg.

With one fork leg holding the wheel in place, you should be able to look down the length of the frame in order to force the rear wheel into lengthwise alignment. While you do this, try not to disturb the horizontal angle of the tacked fork leg, or this will alter your frame-to-ground clearance. With a little effort and careful eying of the frame, you should be able to manipulate the wheel until it is in a perfectly aligned position. Once you have it, just add a small tack weld to the other fork leg, and this will hold the side-to-side alignment in place.

Now take your frame and wheels back to your ground reference point so you can ensure that your frame-to-ground clearance is still correct (between 5 and 6 inches). If your clearance has changed, you should still be able to force the fork legs back into the correct position without disturbing the side-to-side alignment of the rear wheel now that there is a tack weld on both legs.

Once you have the rear wheel aligned, and both fork legs tacked in place, add some real weld to the joints, checking alignment as you go. If you are worried about weld spatter or heat damaging the rear-wheel tire, you can remove the wheel and place a scrap wheel or just a hub into the forks to hold them in place as you weld all around the joint. Once you are done, take a breather, because the hardest part is now over!

ADDING CHAIN STAYS

The last step in completing the basic frame structure is adding the chain stays. As shown in Figure 11-45, an actual set of chain stays cut from a 26- or 27-inch frame are used to form a triangle between the rear fork legs and the rear seat tube. Without these extra pieces, the fork legs would instantly bend if you tried to sit on the frame (you didn't try this, I hope). Because of the triangular structure formed by the chain stays, this becomes the strongest part of the frame.

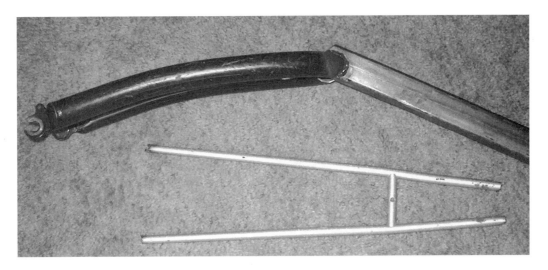

Figure 11-45 Chain stays are added to form the strong rear triangle.

MAKING THE BOTTOM BRACKET

Now you can sit on your frame, since the urge has probably been driving you crazy for the last few hours! Although far from the first test ride, you can see how low and reclined this unit is going to be. Find two old pieces of wood to place between your buttocks and the frame so you can sit right on the Marauder like you will in the final design. Notice that the frame has a small amount of flex, yet still remains very strong and stable. This is the reason why you will not need suspension and can get away with minimal padding on the seat.

While sitting on the frame with a board between you and the tubing, have someone hold one of the crank arms you plan to use in place as if you were making a long pedal stroke. Have him/her mark the place where the crank axel should be on the frame while you do this. If all went well, there should be about two or three inches from this point to the joint between the main frame and the front tube. When you find this point (bottom bracket center), take into consideration the extra inch added from the seat padding, and make sure you do this with your shoes on.

When you find this center point, mark it on the frame, then place a bottom bracket half way up the frame so this line is under its center, and trace the outline. When you have this outline on one side of the frame, use a square to transfer the lines to the other side of the frame, and repeat the process. The areas are then cut and ground out, as shown in Figure 11-46, so that the bottom bracket can be welded to the frame.

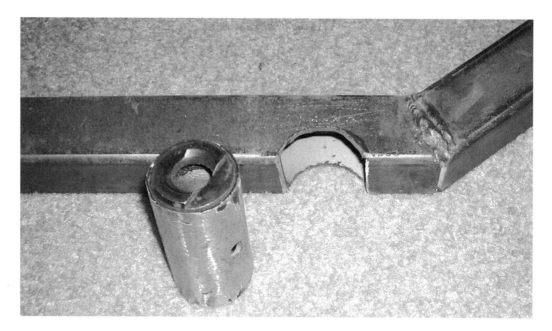

Figure 11-46 Part of the main frame is cut away in order to take the bottom bracket.

While welding the bottom bracket in place, take note of which end is right and left; there is a difference. The left-hand threads should be on the chain-ring side, and the right-hand threads on the non-chain-ring side. Also, place the threaded cups into the bottom bracket before welding, to cut down on the amount of warping that may result from welding heat. As shown in Figure 11-47, weld around both sides and underneath the joint to seal all areas.

The bottom bracket should also be aligned at a perfect 90 degrees to the mainframe tube to ensure proper alignment of the chain ring. You may want to tack weld only one side of the bottom bracket at first, then place a crank arm in the other side so you can compare it's alignment to the side of the frame, making any necessary adjustments. If the chain ring is not aligned with the frame length, you may have chain derailment problems.

THE STEERING SYSTEM

Now, unless your arms are 6 feet long, you are going to need some way to bring the steering closer to your body. This is accomplished by creating an "indirect" steering system. Indirect steering means that you are not directly controlling the front wheel from the head tube. Instead, a

Figure 11-47 The bottom bracket is welded into the cutout in the main tube.

rod with two ball joints at each end transfer steering movement from a remote head tube to the main head tube, almost like the tie rods in a car keep the two front wheels synchronized.

There are four parts that make up the remote steering head (see Figure 11-48)—a head tube, a fork stem (salvaged from the forks used at the rear of the frame), a short piece of tubing of the same type as used for the rest of the frame, and a small control arm that will hold the ball joint.

The fork stem can be trimmed down to about half an inch under the bearing ring where the two fork legs used to be connected, since this area only needs to hold the control arm now. The fork stem shown in Figure 11-48 has already been trimmed. The control arm is just a two-inch by one-inch piece of 1/8-inch plate steel with a hole drilled at one end. This hole is used to bolt on the ball joint. I have used lengths between two and three inches for the control arm, and have found no difference in the feel of the steering, as long as the distance from the center of each fork stem to the bolt hole on the control arm is the same in the front (main steering head) as it is in the back (remote steering head).

When the remote steering head is complete, it will look like the one in Figure 11-49. The angle formed at the mainframe tube and the tube

Figure 11-48 The four parts that make up the remote steering head.

Figure 11-49 The complete remote steering head welded to the main frame.

connecting the remote head tube "E" is 145 degrees. This angle is chosen so that both head tubes are at approximately the same angles in the final design, and it makes the remote steering feel as though it were direct steering. The small piece of tube that joins the remote steering tube to the main frame is placed so that there are 11 inches between it and the seat-back tube. This tube will allow you to install an 11-inch long seat base on the frame (giving you plenty of room to shift around on while riding). The length of this tube is not critical, but it should be long enough so that the fork stem can be turned in a complete circle without the control arm hitting any part of the frame. In my design, I used six inches as the length.

A sharp builder looking at Figure 11-49 will notice that once this remote steering system is welded to the frame, there is no way to completely remove the fork stem without breaking the weld because it cannot be pulled from the bottom without hitting the frame. This has not been a problem, and if you ever have to change the bearings, it can be done by using individual balls (with the retaining ring removed) in the lower half of the unit. Keeping this in mind, make sure that the entire remote steering system is put together before welding it to the frame, or you may find yourself grinding off some freshly made welds.

THE CONTROL ARM

The control arm on the remote steering head must connect to the front forks through two ball joints and a rod, so a similar control arm must be placed on the front forks. As shown in Figure 11-50, this control arm is made in the same fashion as the control arm for the remote steering fork stem. The only difference is that it is made shorter so that the distance from the center of the fork stems to the center of the control arm holes is the same for the front and back. Depending on your front fork's crown width, this may vary, so choose the appropriate length. Also, this control arm is welded to the left side of the fork (looking at it from the back) so that the rod connecting the two will be on the opposite side of the bike from the chain ring. This side of the bike is chosen for the connecting rod because there is more room—the chain, chain ring, and front derailleur are all on the other side of the frame.

To connect the two control arms together, use two small ball joints and a steel rod (see Figure 11-51). Small ball joints like these can be found at many auto supply or hardware stores; they are sometimes used to link such things as gas pedals or windshield wiper control arms. If you cannot find small ball joints, the type used on a snowmobile steering linkage will also work, although it will be larger than nec-

Figure 11-50   The control arm is welded to the crown of the front forks.

Figure 11-51   Control rod and two ball joints connect the two control arms together.

essary. The rod used in between the ball joints can be either a solid steel rod of at least 3/8″ thickness, or a strong hollow tube similar to the size of tubing used for the chain stays on a bicycle frame.

The length of the control rod is determined by the spacing between the two control arms and ball joint ends while both control arms are set at 90 degrees to the frame (steering straight ahead). As shown in Figure 11-52, both control arms extend outwards from the frame when the front forks are in the neutral position. If your ball joints did not come with threaded ends, or you just want to make sure they never move, then weld them directly to each end of the control rod as I did, taking care not to overheat the ball joints by welding only a little bit at a time.

You could use the threaded ends of the ball joints if you also cut threads into the control rod, but this is really not necessary, since adjustment of the steering alignment will be done by placing the gooseneck into the remote steering tube at the desired angle, not by adjusting the position of the ball joints on the control rod like you would to align a car's steering.

THE SEAT BASE

Now that the remote steering system is complete, you will know how large to make the base of the seat—the distance between the tube that holds the remote head tube to the frame and the rear seat tube. Before you can mount a plywood seat base, six plates of thin steel are cut and welded to the frame (see Figure 11-53). These plates are made from 1/8″-thick plate and have small holes drilled to allow wood screws to be

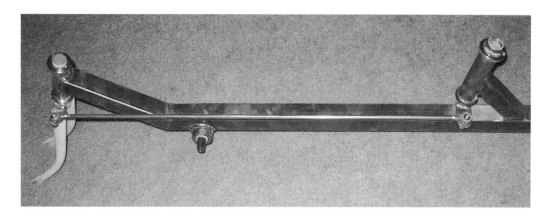

Figure 11-52 The control arms will be at 90 degrees as the bike steers straight ahead.

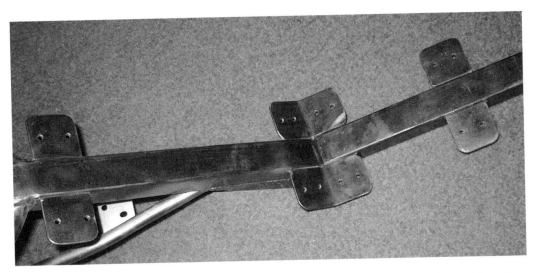

Figure 11-53 Metal plates are welded to the frame in order to hold down the seat boards.

used. Notice that the center plates are bent in the middle in order to conform to the frame and also join the two seat boards together. Once the plates are welded to the frame, you can cut out seat boards from half-inch plywood to whatever shape and width you like. I found a width of 11 inches to be just right, but feel free to experiment. You will need to screw down the seat boards now, since the next step requires them to be in place.

REAR WHEEL, DERAILLEUR, CRANKSET, AND CHAINS

Once you have your seat boards in place, put the rear wheel, rear derailleur, and crankset on the bike. You will now need to join about two-and-a-half regular bicycle chains together in order to make one long enough to reach the entire length of the Marauder. As you connect the chain, notice how it rubs under the front of the seat board. Do not worry, as this will all be fixed by the addition of a chain guide pulley.

As shown in Figure 11-54, a large Teflon or plastic pulley is placed just under the back of the seat in order to make sure the drive chain gets routed under the seat without rubbing on any part of the seat board or frame. This pulley is held in place by a bolt that is welded to the side of the gusset between the main frame tube and seat back tube. The pulley in my design is six inches in diameter and has a built-in ball bearing. This type of pulley is very common in gym

Figure 11-54 The chain is routed under the seat by the large chain guide pulley.

equipment that uses a cable to lift a stack of weights and can be bought from a fitness supply store. I chose a large-diameter pulley in order to keep the friction from bending the chain line down to a minimum. I have tried smaller (2 to 3 inch) pulleys and noticed they seem to rob a little bit of your energy after a while. It's surprising that many commercially available low racers end up using cheap, small-diameter soft rubber chain guides in their designs despite this. With your chain connected, just hold the pulley in place until the chain just makes it under the seat; there's no need to go any further than that. When you get this point, mark it with a marker or pencil through the bearing hole in the pulley, then weld the appropriate-size bolt to that spot.

MAKING HANDLEBARS

Handlebars are made by extending the shaft of a gooseneck up to the appropriate height, then welding a shortened set of curved handlebars to the top of this extension. Let's start by making the extension for the gooseneck. This way, you can sit in the seat and decide how tall you want to make your handlebars.

As shown in Figure 11-55, a gooseneck is cut so only the stem remains and a piece of ¾-inch tubing is welded off the side up to the required height to hold the handlebars. Because you don't know this height, cut the tube being used as the extension to at least 20 inches. Once the two parts are joined (see Figure 11-55), place the neck into

Figure 11-55 An extension tube is welded to the side of a modified gooseneck.

the remote steer tube, sit on the seat, and pedal the cranks, taking note of how high the handlebars would need to be in order to clear the highest point of your legs.

Once you have determined how high the handlebars need to be, cut the extension tube at the appropriate height, taking into consideration seat padding, if you haven't added it yet. From here, a curved set of handlebars from a mountain bike are cut short (13 inches wide) and welded directly to the top of the newly added extension tube, as shown in Figure 11-56. The reason for the curve is so that your wrists will be in a natural, relaxed position while you ride. A completely straight set of handlebars would feel uncomfortable after awhile. As for the narrow width of the handlebars, this is done to keep your arms as close to the sides of your body as possible, to increase your aerodynamic position. If your handlebars were as wide as the ones found on a regular bike, you would collect wind at your sides, and this would slow you down slightly, which is something you don't want for this type of racing machine. Because of the narrow width of the

Figure 11-56 A short curved set of handlebars are welded to the extension tube.

handlebars, shifters where mounted on the extension tube to save room.

DISC BRAKE

Because the Marauder can easily reach speeds well beyond what you may be used to on an upright bike, good brakes are not an option—they are a requirement. You could get away with a decent set of caliper brakes mounted to the rear forks, but I have tried this and found that you will wear those little rubber pads out every few days if you ride hard, and the stopping power produced by these brakes is less than adequate.

An alternative to caliper brakes is the disc brake. This brake is gaining in popularity and can be purchased as a kit (brake, cable, and lever) for a reasonable price. To mount a disc brake, you will need two things—a rear wheel with a hub that supports the mounting of the actual disc, and a plate to mount the braking system on the rear forks. This plate is made from a piece of 3/32-inch steel plate formed to whatever shape is required to hold the brake to the forks. As shown in Figure 11-57, this plate is mounted just above the dropouts on the rear forks.

In my design, I utilized a fully hydraulic disc brake system. This system is not only smooth and silent, but has enough power to lock up the

Figure 11-57   A plate is made to mount the disk brake on the rear forks.

wheel at any speed, even on clean, dry pavement. I originally used a caliper brake in the original design but after melting down the rubber pads on a hill at more than 53 miles per hour, I would never go without the disc brake again!

RIDING THE MARAUDER

Well, what are you waiting for? Add the rest of the components onto the bike and head out for a test ride! Learning to ride the Marauder is a simple matter of losing the "shakes." For some reason, everyone who learns to ride the bike tends to wobble around a little for the first hour or two before they get used to its handling. Once you get used to the feel of being so low to the ground, you will really enjoy the comfortable and exciting ride that this bike can deliver, and you may put your upright bicycle away for good.

Well, I hope you enjoy the Marauder as much as I do. The design is more complicated that the Kool Kat and Coyote, but you'll soon appreciate that all of the measurements, details, and patience are necessary to make a lean, mean, racing machine. The effort is definitely worth it! The first time I rode the Marauder around the block, I was addicted,

Figure 11-58 The Marauder is so low to the ground that you just put your hand down to stop.

not only to the amazing handling, blasting into a high speed corner, but also to the absolutely comfortable feel of the bike even after hours of hard riding. I doubt that I will ever again ride an upright bike after enjoying such a great recumbent vehicle.

Now that you're pumped up from your low-racer bikes, let's have some fun in the next chapter with a couple of easy-to-build creations, the Spincycle and the Sidewinder.

12

UNCLASSIFIED ROLLING OBJECTS

The Spincycle is a fun little trike that allows the rider to perform all kinds of crazy spinouts and trick maneuvers. The idea for the Spincycle came from my friend, Troy Way, during the cold winter months when all you can do is design rather than build custom bikes if you're like me and don't have a heated garage.

Troy's idea was to have a bike or trike with free-steering caster wheels that will allow the vehicle to spin and slide all over the road with very little input from the drive. The second he told me about this idea, I knew it would be fun.

First, we needed some type of caster wheel that could rotate in a 360 degree circle. Immediately, shopping cart front caster wheels came to mind. These wheels are tough, small, and easy to work with, so they were chosen as the rear wheels. Now we had to figure out a way to drive the unit. The trick was to get power to one of the wheels while the other wheels could rotate freely in all directions, so driving the rear caster wheels was out of the question. Logic would dictate that there had to be some type of front-wheel drive on the trike if caster wheels were to be used as the rear wheels. We opted for a front-wheel-drive design that I had tried on a fairly successful low-racer called the Coyote that I had built a few years back. The Coyote had a bottom bracket mounted off the front forks so that the front wheel was placed between the rider's legs. The chain made a short run to the front wheel and the unit was steered by turning the front forks and crankset all at the same time.

Now that we had a transmission system and rear-end design, the rest of the plan was simple. Place a seat in between the front wheel and the rear casters as low to the ground as possible so the trike would not be able to flip during a 360 spin out. In Troy's original prototype, the frame was made of just two 1.5-inch square tubes welded into a "T" shape with the two rear caster wheels mounted to a 24-inch bar at the rear. The first test ride was great! The unit worked exactly like Troy envisioned, spinning and twisting all over the place. You could actually control the trike if you wanted to, then, with a simple flick of the steering, send it into a wild uncontrollable spin!

Even though I could barely reach the pedals on Troy's prototype, I had amazing amounts of fun tricking out on the trike, so much that I just had to make one as well. Here is how it was done.

Building the Front End

We will start with the most complicated part of the Spincycle—the front end. You will need a good 20-inch BMX rear wheel, a bottom bracket with the chain stays attached, and a set of 20-inch BMX forks like the ones in Figure 12-1. This vehicle will be spinning wildly all over the road, so the stronger the parts, the better.

A wheel with 36 or 48 spokes will be good, but try to avoid a wheel with fewer than 36 spokes. As for the forks, the type with solid one inch or larger legs will be good; avoid the tapered style, as they will not work. You can use either a freewheel or a coaster brake hub in your wheel since brakes are optional here. Troy's original design has a freewheel hub, and mine has a track cog (direct drive) so I could ride the trike in reverse as well.

Figure 12-1 A 20-inch rear wheel, BMX forks, and bottom bracket are used in the front end.

The only requirement for the bottom bracket and chain stays is that they must span the wheel when the ends are placed on the front forks (see Figure 12-2). They are placed as close to the end of the forks as possible to allow the chain to reach the sprocket without rubbing. If you imagine the unit with the chain in place, the chain stays will be in between the top and bottom run of the chain, just as they were on the donor bike. Bottom bracket type is not important; one-piece or three-piece cranksets will both work just fine.

Before you make any welds, you should widen the front forks to accommodate the rear wheel. The rear hub is wider than a front hub, so you will need to stretch the forks apart about an inch or so. Try putting the rear wheel in place to see how much you will need to widen the fork legs. The easiest way to do this is to stand with one foot on a fork leg and pull up on the other. If you don't have the strength to do this, find someone who does. Take it easy when bending the fork legs, and

Figure 12-2   The chain stays are placed at the bottom of the front forks.

do it in little steps until the rear wheel will fit into the dropouts. Strong aren't they?

Once your fork legs and chain stays are wide enough to take a rear wheel, weld the chain stays to the fork ends at about a 90 to 95 degree angle, as shown in Figure 12-3. Make sure they are aligned properly by placing a wheel in the forks to check for tire rubbing. Tire rubbing is bad! If you did not cut each leg of the chain stay to the same length, your tire will hit one of the sides. The angle of the chain stays is also important. If the bottom bracket is too close to the ground, the pedals will hit during a turn. Take a look ahead at the final trike (Figure 12-13) to get an idea of where the bottom bracket will be in the completed design.

There is only one more part to the front end—a tube that runs from the top of the forks to the bottom bracket. I will refer to this tube as the boom from now on. In Troy's prototype, the boom is simply a straight piece of tubing. The advantage to this design is simplicity, and it works very well. The disadvantage is that your fork angle and bot-

Figure 12-3   Weld the chain stays at an angle slightly more than 90 degrees to the front forks.

tom bracket height have to be a little higher. In my design, I used a piece of curved one-inch electrical conduit that follows the curve of the front wheel so I could have a lower bottom bracket.

Bending a piece of one-inch tube is easy if you have one of those manual electrical conduit tube benders, or know an electrician who has one. We of course, did not have access to such fine tools at the time, and in the true spirit of Atomic Zombie, bent the tube around a rusty car rim by pushing down on each end of a ten-foot-long piece. Although this crude method takes a lot of grunting and patience, it will produce an acceptable bend like the one in Figure 12-4. The tube should meet the top of the forks and the bottom bracket while giving adequate clearance between the top of the tube and the tire. If you bend your tube on an angle that's too steep, it will rub on the tire, but this can be easily fixed by just bending the tube in the opposite direction until it fits. If you have no luck at all bending the tube but want the lower bottom bracket, you could just weld two or three straight pieces together at slight angles to follow the contour of the wheel. Once you do have the boom in place and have made sure the front wheel will not rub, weld it all together.

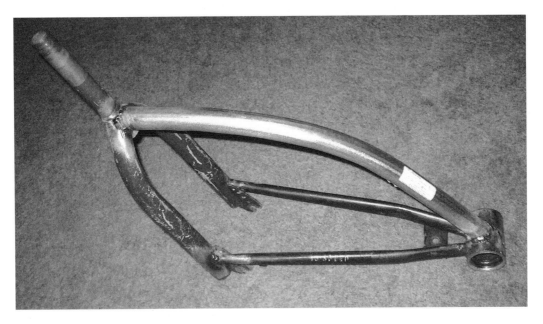

Figure 12-4 The boom is made by bending a one-inch piece of electrical conduit.

Building the Rear End

Building the rear end of the trike is a lot less work than building the front. Simply cut out a 24-inch piece if 1½-inch square tubing like the one in Figure 12-5. A 1/8- or 3/32-inch walled tube will do the trick, but avoid anything thinner, as strength is more important than weight here. You could get away with a round tube for the rear but, as you will see, connecting the shopping cart casters will be more difficult. The caster wheels will be bolted into the tube by drilling a hole and placing the nut inside the tube. This is the original design from Troy's prototype, and it worked perfectly, so why change a good thing?

Find a drill bit slightly larger than the bolt on the caster wheels and drill a hole about one inch from either end of the tube. The caster wheel will just be bolted to the tube with a nut inside the tube (see Figure 12-6). Don't drill the hole too far away from the end of the tube, or you will not be able to get your wrench inside to tighten the nut.

Shopping cart casters are in abundant supply in many places. There are always plenty of bent ones out at the dump (municipal landfill site) or the scrap yard with perfect caster wheels. You could also go to a grocery store and ask the manager if they have any scrap carts you could have. Some stores may be glad to get rid of them.

Figure 12-5 Rear part of the frame is just a two-foot piece of square tube.

Figure 12-6 Shopping cart caster wheels are bolted so that the nut is inside the square tube.

To remove the front caster wheels from a shopping cart, just take off the one nut and they will drop right off. If you have a choice of wheels, take the ones that have the best bearings. If the wheel flops around too much, it will flutter and vibrate as you ride the trike in a straight line at higher speeds.

The rest of the frame is made of the same square tubing and a head tube that matches the front forks you are using. The head tube will be welded to a tube so it is at a 45 degree angle to the ground. Keep in mind that none of these angles are absolutely critical, and the final design is really based on your leg length and donor parts. The tube that carries the head tube should be ground so that there is a good weldable joint between the two, as shown in Figure 12-7. The better the joint, the easier it will be to weld the round tube to the square tube.

The main frame can be made of either a single tube, as in Troy's prototype (see Figure 12-16), or from three separate tubes laid out like the ones in Figure 12-8. Either method will work fine, but the three-piece frame will allow for a lower center of gravity. The actual lengths and angles of the tubes will be dependant upon your leg length and head tube angle. There are no magic formulas here; just cut the tubes to fit. The goal here is the have the bottom tube (horizontal tube) about an

Figure 12-7   The head tube is welded to the square frame tube.

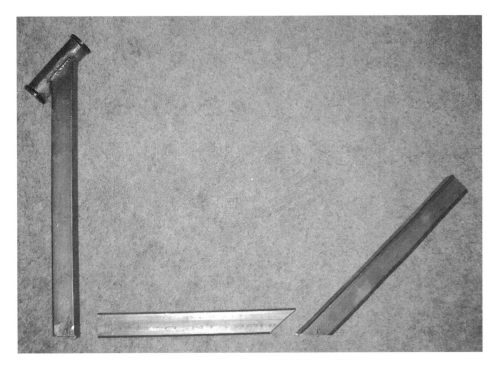

Figure 12-8 The three-piece frame layout allows for a very low trike.

inch from the ground and to have the backrest tube set up so that you can reach the pedals when you sit in the frame.

To get all of the frame tube lengths and angles without spending hours making calculations, just lay out the frame on the ground and sit in it. Start by cutting a 16-inch piece for the backrest tube (this is the tube that holds the back of the seat). Lean the backrest tube up against the wall so that it is at about 45 degrees. Now sit on the floor so you are resting against the backrest tube and have someone hold the front end (with pedals attached) so it is in the correct place for your leg length.

Remember that the angle of the head tube should also be at about 40 degrees. When all seems to fit, just measure the required length of the bottom tube (horizontal tube). The bottom tube will carry the bottom of the seat, and must be at least two inches from the ground in the final design for clearance.

I put a 2-inch piece of tube under the frame when I laid it out, as shown in Figure 12-9. Also, notice the minimal clearance between the tire and the down tube (which joins the head tube to the bottom tube). When you are first making the main frame, tack weld it all together

Figure 12-9   The main frame should fit the rider and have a two-inch clearance underneath.

just in case something does not work out. Put in the front end, and check the angles and clearances before you do the final welding.

Once your main frame is made, all that is left is to join the rear bar (holds the caster wheels) to the main frame. A square notch is cut from the rear tube so that the caster bar can be welded directly to it (see Figure 12-10). Before you make this cut, there are a few things to keep in mind.

This notch cut must be made so that the lower frame maintains that 2-inch clearance between the horizontal seat tube and the road. Also, the caster bar should have a slight forward angle of about 15 degrees as well, as shown in Figure 12-11. This slight forward angle allows the casters to fall into a straight line while you are moving in the forward direction or standing still. If you did not have this slight forward angle, the trike would be totally out of control at all times, and even the slightest turn would send you into a wild spin. With this angle added, you can actually control the trike quite well at any speed and still send it into a "fish tail" turn or full spin out by leaning into a turn. This forward angle is the same principal that allows you to ride with no hands on a regular bicycle (the front wheel tries to fall into a straight line).

Those strange square blocks with the hole drilled through them are my last-minute modifications to the original design. I decided it would be cool to have the ability to throw on a rear set of 20-inch wheels al-

Figure 12-10 A notch is cut from the rear of the main frame to carry the caster bar.

lowing the trike to convert from caster-wheel mode to full-out racing mode. To do this, you will need a set of two front BMX-style wheels with those hefty 14 mm axles. I will explain how this works later on, in case you want to add this conversion option to your trike.

Once your rear caster bar is in place, your frame is almost complete and should look something like the one shown in Figure 12-12. You will need to add some small pieces of flat bar or plate to allow the bolting on of your seat.

I cut eight 1/8-thick pieces of one inch by one inch flat bar, welded them to the frame and then drilled small holes through them. Your seat can be made from just about anything. Troy used a nice cushy office chair we found at the scrap yard, but since there was only one, I just used two pieces of half-inch thick plywood. It's not like you're going to be cycling across the country on this thing, so don't worry too much about the seat.

Now you can paint your high-performance Spincycle and then add the final bits and pieces. Cut a chain to fit, and find a set of handlebars

Figure 12-11 Rear caster bar is welded in place at a 15 degree forward angle.

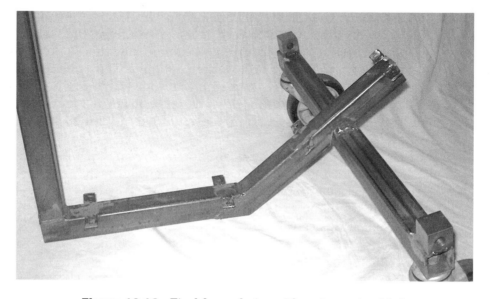

Figure 12-12 Final frame design with seat mounts added.

that do not interfere with your knees as you pedal. Small kid's BMX handlebars worked well on both of our designs despite the differences. Notice how low and laid back the final product is in Figure 12-13. When I am sitting on the seat, I actually look through the handlebars! Now that's a low bike. Driving over a curb is interesting due to the low frame, and creates a few cool sparks at night if I'm going fast enough!

CONVERTING FROM CASTER TO RACING MODE

Here is an explanation about the conversion option. If you would like to be able to transform your trike into a fast and streetworthy machine, then you will need to find a set of BMX front wheels with 14 mm axles like the ones shown in Figure 12-14. Do not even attempt to use a set of wheels with anything less than a 14 mm axle, or the axles will instantly bend when going around the first corner. The 14 mm axles are also a lot longer than the small type so they can carry those stunt pegs you see on the freestyle bikes. As you can see in Figure 12-14, I

Figure 12-13 The completed and painted Spincycle shown in "caster" mode.

Figure 12-14 BMX wheels with 14 mm axles can be bolted to the rear of the Spincycle.

have moved the axles so that all of the extra length is on one side, and this is what will be bolted to those mysterious cubes welded to above the caster wheels.

When the 20-inch wheels are bolted to the steel bocks, the casters are lifted slightly off the ground, transforming the Spincycle from "caster" mode to "racing" mode. Although you will not be able to do any spins or crazy stunts, you will be able to go really fast! With a larger front chain ring added, you will be able to give any upright bicycle a real good run. The trike is so low and reclined that it is capable of some real speed. When I first tried the trike in racing mode, I was very impressed with the way it handled and the speed it was capable of. I may actually build a new unit from lighter materials, with a full set of gears and a better seat, designed to be used as a racing unit.

When you attach the 20-inch rear wheels, a rubber tie strap is placed between the caster wheels so they do not bounce around or hit against the spokes.

Riding the Spincycle is very easy and you will be able to perform stunts that would be totally impossible on any other type of vehicle. When you get some speed going and begin to turn, the rear casters will also want to turn, putting you into a slide similar to that of a car on ice. At this point, you can just turn or accelerate out of the slide to keep moving in a straight line or turn with the slide to go into an uncontrollable 360 spin. When you start spinning, it's like a type of carnival ride that you have little control over! If you practice your spins, you can actually learn to take back control of the trike and keep moving in a straight line after any number of 360's.

Having two or more Spincycles is more fun than I can put into words. Racing around a track with a few obstacles to avoid while you try to grab your buddy's seat, throwing his unit into a wild spin is just too much fun. Eventually, we hope to produce several units and have some crazy Spincycle racing events. Troy and I practiced our controlled

Figure 12-15 With the rear 20-inch wheels attached, the Spincycle is a speedy low-racer.

spins to the point where we could do synchronized stunt driving that would draw a crowd anyplace we did them. It's impossible not to have fun on the Spincycle. Everyone who tries it does not want to get off after a few minutes.

In Figure 12-16, you can see how different our designs are, yet both work equally as well. As long as you design your trike so the rear caster bar has a slight forward angle, and you can reach the pedals, it will work just fine.

After a few hours of spinning ourselves into a dizzying sickness, we came up with several games. One was the circular obstacle race. Just try to beat your opponent around a circular track with several obstacles along the way such as tires or pylons, and use any means you can to send your opponent out of control. A quick grab and pull on his trike or cutting him off into a turn is usually all it takes. Another game is similar to tag. You can place a flag or ball on the rear of each trike and try to grab it from your opponents as you race around an open area. I thought it might be fun to knock a ball off your opponents hat with a padded broom handle or stick—kind of like "bicycle jousting." The possibilities are endless, I'm sure. In Figure 12-17, I am about to slam into Troy's Spincycle as he slides out of control in a game of "chicken." Ouch!

Even in racing mode, when the two rear 20-inch wheels are connected, it is impossible to flip due to the low center of gravity. If you plan to

Figure 12-16   Two Spincycles ready to duel. Notice Troy's high-quality office chair seat.

Figure 12-17 A view from the helm during a game of chicken. Troy slides out of the way.

Figure 12-18 Devon "Road Kill" Graham test drives the Spincycle in "racing" mode.

ride the trike like this in traffic, you will have to add some type of brake, as the unit can get up to some dangerous speeds. A cable brake to the front wheel would provide plenty of stopping power since most of the weight is over the wheel. It would also be easy to add a gear cluster and derailleur to the front wheel as well as a multiple chain ring crankset and front derailleur. Feel free to experiment and have some fun with the design.

Overall, this was a worthwhile project and never fails to give hours of enjoyment every time I bring it out. If you just want a fun vehicle to play around on, this is definitely the one!

The Sidewinder

Just like the Spincycle, The Sidewinder is a trick bike capable of stunt work that a normal bicycle could never perform. The key to this bike's agility is the ability for the frame to basically fold in half while you are riding it. As the bike folds, the front wheel is brought closer to the rear wheel, resulting in a very short wheelbase. The shorter the wheelbase of a bicycle, the tighter the turning circle that is possible. When the Sidewinder is folded, it is capable of turning a circle in a width of less than two feet! Try that on your regular bike!

Building the Sidewinder is not difficult and needs only one donor bicycle and a few short lengths of electrical conduit. The plan will be to make a long frame that can fold in the middle, just ahead of your knees. I have seen other folding bikes on the Internet, and found one called the "swing bike" that was actually produced in the 1980s, but found that its design would make it very difficult to pedal while in the "folded" mode. The problem was the placement of the hinge that allowed the bike to fold or swing. If the hinge is too close to the seat tube, your legs will just hit the top tube as it begins to fold, making it hard to pedal and restricting the amount of fold. To get around this, I moved the pivot point far enough ahead so the bike can fold almost in half while you are still able to pedal. This ability to fold into a smaller package really increases the amount of fun you can have on the bike.

CUTTING AND ARRANGING THE FRAME

Start with a frame of any size and shape you have laying around. I choose a simple ladies' mountain bike frame for my base frame. Cut the frame in half where the top tube and down tube meet the seat tube and bottom bracket, as shown in Figure 12-19. You can also grind the leftover stubs from the seat tube and bottom bracket at this time.

The next step is to lay the frame and forks (with wheels and crank arms attached) on the floor so you can get an idea of the overall position and angle of the tubes. The idea is to make a pivot out of the original head tube so that the frame can fold just ahead of where your knees would come up as if you were pedaling the bike. After the pivot, the frame is made to stretch about 14 inches in length. The majority of angles and lengths are not critical, but there are two things to keep in mind—design the unit so the pedals do not hit the ground, and make sure your knees will not reach past the pivot point. Other than that, you can do whatever you like.

Notice in Figure 12-20 that the original front of the frame (with the head tube) has been shortened and turned upside down. Turning this piece upside down allows the head tube angle to remain at the same angle it was at originally before the tubes were shortened. This angle helps the bike turn into a short circle, as it lifts the frame during a fold maneuver. Remember to keep track of the pedal clearance as you lay out your frame tubes for cutting, making sure there is at least three inches between the bottom of the pedal and the road. It helps to lay down a long stick or string to simulate the road as I did in Figure 12-20.

To get the proper length for placement of the pivot point (old head tube), sit on a regular bicycle (if you have such a thing) and pedal in re-

Figure 12-19   Cut the donor frame in half at the seat tube.

Figure 12-20 Lay the frame and tubes on the floor so you can see the overall design.

verse, measuring the farthest point where your knee passes the main tube. Add two inches to this distance and this is where the center of the pivot point should be. When you have this distance and have cut the tubes, weld the front of the frame back in place, as shown in Figure 12-21. Remember to provide pedal-to-ground clearance before you attack the tubes with your hacksaw. Try to get the two halves aligned as they are welded in place, or your bike will have a hard time going in a straight line, something you may want to do once in a while.

The next step is to make the bottom half of the pivot part of the frame. This piece is made by welding a one-inch tube or piece of electrical conduit to the bottom of a cut-off set of forks, then to a new head tube. As shown in Figure 12-22, the bottom pivot tube is welded to what is left over at the bottom of a pair of forks with the legs cut off and ground. Remember to lay out the frame on the ground like it was in Figure 12-20, so that you can keep track of the ground clearance as you decide where to cut the tubes.

The length of the pivoting part of the frame is not critical, but do make it at least 14 inches from the pivot point to the front head tube, so that the front wheel will not hit the pedals during a tight folding turn. I made my front frame length 16 inches from the head tube center to the pivot point center, and this worked out well.

Figure 12-21   The front of the frame is shortened, inverted, and welded back in place.

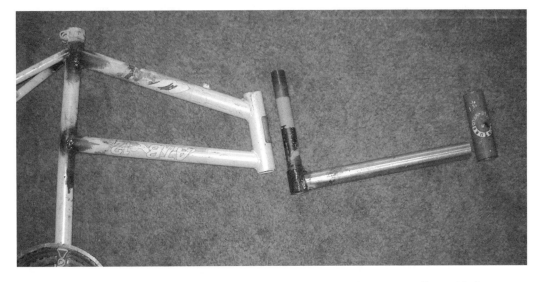

Figure 12-22   The pivoting part of the frame is made from a cut-off set of forks.

WELDING THE PIECES

Once you have chosen the proper angle and length for the bottom part of the pivot frame tube, weld it to the bottom of the fork stub, as shown in Figure 12-23. This part of the forks is the strongest part, as it has an extra-thick base where the fork legs were connected. It's a good idea to remove the bearing ring before you weld, just in case you slip and hit that area with the welding rod. When you are done welding the pieces together, just tap the ring back in place with a hammer.

Once the bottom pivot tube is welded in place and the new head tube is welded to the end of it, the top of the pivot frame will be made from another one-inch piece of tubing and an old gooseneck. Place the bearings and rings into the pivoting head tube so you can measure the length of tube needed to run from the gooseneck (in the pivoting head tube) to the front head tube (see Figure 12-24). Try to get this tube at the correct length before you weld, so you do not have to fill up a big gap at the joint.

Figure 12-23 Weld the bottom of the pivot frame tube to the bottom of the cut-off forks.

Figure 12-24 One last piece of tube and a gooseneck finishes the pivoting part of the frame.

Although this design is very simple and effective, it does have one disadvantage—you cannot ever take the front pivoting part of the frame apart once both tubes are welded. Of course, this is just a fun bike, and the chances that you will ever need to change the bearings in the pivoting head tube due to wear are almost zero. In fact, I can't remember ever having to replace any head tube bearings due to wear!

Before you weld the final top pivot tube in place, make sure the bearings are greased and all the rings are put in place, as this will be your last chance to do so. As you can see in Figure 12-25, the final front part of the frame will never be able to be removed without cutting the top tube. If you really want to be able to take the two parts of the frame apart, then you could fashion some type of connection from the cut-off gooseneck to the front of the head tube to allow this tube to be removed, but this is really not necessary.

ADDING THE CHAIN, SEAT, AND HANDLEBARS

Now, with the frame completed, you can add your chain, seat, and handlebars to give the Sidewinder a test ride. Notice that in my freshly

Figure 12-25   The completed front pivoting part of the frame has been welded.

finished prototype (see Figure 12-26), the pedals have adequate ground clearance, and the head tube and pivot point tube are at roughly the same angle. If you did not plan ahead for pedal clearance, and your pedals scrape the ground, you have only two options: cut and reweld parts of the frame or add an extended set of front forks to raise up the bike.

After a few test rides, I decided that a banana seat would be better as it allows the rider to move into different positions during those wild snake-like twists and turns. Also, a banana seat is more forgiving to riders of different leg lengths, so riders of all ages could give the bike a go. The unit handled very well, and was able to perform the shortest turning radius of any bike I have ever tried. The banana seat and fresh coat of paint really improved the look of the final bike (see Figure 12-27), and, overall, it has turned out to be a complete success.

Notice how short the wheelbase becomes as the bike is folded even half-way as in Figure 12-28. It is possible to fold the bike while you are

Figure 12-26   Completed prototype ready for an initial test run.

Figure 12-27   A banana seat is a better option for the Sidewinder.

Figure 12-28 The Sidewinder can be folded so that the front wheel meets the rear wheel.

riding so that the front wheel is actually overlapping the rear wheel during a turn, and this creates some fun maneuvers. Another fun thing to do is rapidly fold the bike from one side to the other while you are riding, producing a snake-like motion as you slither down the street. When you ride up to unsuspecting bystanders, they do not realize the bike is anything out of the ordinary until, all of a sudden, you pull the frame into a fold and perform some impossible circle in front of them. Yes, this bike gets some strange looks!

For fun, set up an almost impossible obstacle course that only a snake could manage to travel through and try to wind your way through it without hitting anything. If you really want to have some fun, challenge your buddy to try it on a regular upright bike! After a few rides, I managed to get my turning circle down to under two feet

Figure 12-29   The Sidewinder can't stop on a dime, but it can turn a circle on one.

(see Figure 12-29), less than a small child's tricycle. Oh, and the Sidewinder can be stored in a small space by just folding it in half!

Both the Spincycle and Sidewinder are examples of the types of creative projects you can create when you inject some ingenuity and imagination into your work. If you don't take your designs from the drawing board out to the workbench, you may be passing up a great opportunity to invent something new or just have a whole lot of fun!

13

FINISHING AND RESTORATION

Painting with a Spray Can

So, you've just put the final weld on your latest creation and took it out for the first successful test ride. Don't fall victim to laziness now and bypass the painting stage! Your cool creation will look so much better with a paint job, any paint job. Even the worst paint job you could possibly do will still look better on your final design than those freshly welded bare and discolored tubes. Painting isn't difficult, and anyone can do a fairly good job with just a can of store-bought spray paint.

The key to getting a nice finish from a spray can is patience and more patience. If you clean and prime the surface, then take your time spraying on two or more light coats, the paint will adhere to a smooth and blemish-free coat every time. Spraying the paint on too quickly will create runs and drips in the final product as it rolls off the surface.

Priming the surface may not always be necessary. If there is an existing coat of paint on the metal, a light sanding to take out the gloss and bumps is usually enough to make the paint stick. If the majority of your frame is made from new metal, then an initial coat of metal primer will make the final finish a lot more resistant to chips and scratches.

Clean all dirty or oily surfaces with paint thinner or solvent, then wipe the area down thoroughly with a rag. A light sanding with a fine emery cloth or sanding pad will also help paint stick to smooth galvanized surfaces such as new electrical conduit and exhaust tubing.

The first coat that you apply should be so light that it only slightly changes the color of the surface. This initial coat will just set up the surface so that the other coats will stick to it without causing runs or bubbles. Use short, quick strokes along the length of a tube to avoid heavy paint buildup or runs. Always shake the spray can for a good two or three minutes before you start painting, and shake it again after about ten strokes. If you notice a buildup of paint near the exit of the spray nozzle, wipe it off with a cloth so it doesn't spit a blob of paint onto your work.

Refer to Figure 13-1. The top tube was painted using a heavy first coat in an attempt to get good coverage right at the beginning. The result is a fairly good covering of paint at the top with massive drips and runs along the sides where the paint rolled off the tube. This is what happens when you try to rush a paint job, and it's a telltale sign of an amateur who doesn't have patience. The bottom tube has such a slight dusting of paint that it almost looks bare. This is the proper method of applying the very first coat. This initial coat will dry fairly quickly, allowing you to turn the frame to reach the areas that were hidden or hard to reach in the beginning. Repeat the process until the entire frame has a light dusting of paint or primer in an equal coating.

Once the initial light coat has dried (see the paint directions on the can for drying time), you can add the second and final coat. The second coat will begin to cover the metal but, once again, don't overdo it or you will end up with a lot of runs and bubbles. Expect to do a third coat and even a fourth if you want a really good paint job. In Figure 13-2, you can see that the second coat covered most of the tube without running or creating bubbles due to the careful application of the first coat. Once the third coat has been added, the tube will look as good as if it were done in a professional paint shop.

Paint is also subject to something called "curing time." This is not the same as drying time. Your final coat may appear to be dry after resting in your shop overnight, but you will not be able to handle the

Figure 13-1 Too much paint will cause runs and streaks.

outside work for quite some time. Sure, as long as you are careful not to ding or scratch the surface you will be OK, but even the slightest scratch will easily damage a fresh coat of paint. Curing time for paint can vary greatly, but I have found that most spray paint will be fully cured within a few days of the final application. Fully cured paint will not scratch off or chip easily, but uncured paint can be picked off with

Figure 13-2 Properly applied paint will not have runs or bubbles on the surface.

as little as a light scratch with your fingernail. If you want your paint to last, take it easy for the first few days.

Restoring Chrome from Rust

Many of the bicycle parts you will be using will have some amount of surface rust and caked-on grease unless, of course, you have an unlimited budget and only work with new parts. A lot of this rust is only on the surface and can be buffed right off with little effort using simple methods. Of course, if the surface is so rusted that there are pits or holes, then no amount of effort will bring it back to life, not even paint.

If the rust isn't too thick, you can usually just buff it off by hand using a rag or cloth. Wipe the area clean with solvent first, to remove any dirt and grease. In the top part of Figure 13-3, we see a rim that

Figure 13-3 Surface rust can usually be removed from chrome with little effort.

has blotchy surface rust over the entire area. Leaving a bicycle out in the rain or storing it in a damp area usually causes this type of rust. In the bottom part, the rust is all gone and the brilliant shine of the chrome has been brought back to life. Removing this rust was done by hand with only a rag, and took very little time and effort.

There are various other tools available for your arsenal in the battle with rust. Wire brushes, grinder brushes, buffing wheels, and fine-grit sandpaper are just a few of them. The grinder brush shown in Figure 13-4 should only be used on chrome as a last resort, as it is a powerful tool that will leave fine scratches as it cleans. The grinder brush will also remove hard rust from bare steel and any amount of paint. The hand brush is much less abrasive, and will help to loosen dirt as well.

To resurrect the shine from dull chrome, a buffing wheel can be used. This is a circular attachment that fits onto your grinder or drill and gives the effect of a spinning rag. If you had to resort to using a grinder brush, a buffing wheel could be used after to buff out the fine scratches.

The real question is how much time and effort do you want to spend trying to restore a part? If it just has a light covering of surface rust, then you will have no problem at all. If your part is deeply rusted or

Figure 13-4 A grinder brush, hand brush, or buffing wheel can be used to remove light rust.

pitted, then it may not be worth the time, unless it is some rare part that you cannot get new. Sometimes, using a little silver paint ("gypsy chrome") may be the only solution.

Fixing Dents and Bends

Bent rims, dented fenders, and bent forks and stays are some of the things you may have to deal with when working with used or salvaged parts. Sometimes, these problems are the reason you actually got the part for free. Most of these problems can be corrected with a little time and effort as long as they are not too severe. If your rim looks more like a soft taco, then you should just cut out all of the spokes to salvage the hub and throw the rest away.

Let's start with rims. If the rim is not too far gone, usually just adjusting a few spokes is all that is needed to bring it back to life. Give the wheel a spin while it is mounted on the bike and give it the "warpage" test. If any part of the rim or tire is so far gone that it hits any part of the frame, then you should just give up right now. If there are only slight deviations in the rim, then you should be able to get it back to normal, or something fairly close to it.

Remove the tire and liner from the rim, then place it back into place with the bicycle turned upside down so you can spin the rim around. Start by locating the worst part of the rim, the part that wanders farthest off center. The plan is to tighten the two closest spokes on the opposite side of the bend about a half turn to begin pulling the bend out of the rim.

Refer to Figure 13-5 and you will see that by tightening a spoke, it will have a tendency to pull the rim in the direction of the spoke angle. Find the two closest spokes to the bend (these will usually be crossed over each other) and start with them. Give the rim a spin to gauge how much correcting has been produced and continue the process. You may need to loosen a few spokes on the opposite side of the bend as well, depending on the amount of bend in the rim.

This process will usually work on most slight bends, but will not take out sharp or severe bends like the kind you inflict on a rim by hitting a curb or steep pothole. As you learn which spokes to tighten for which part of the bend, this process will become easy to get right. Once you do get the rim back into decent shape, make sure the rest of the spokes are fairly tight as well, since these radical adjustments may have loosened some of the other ones.

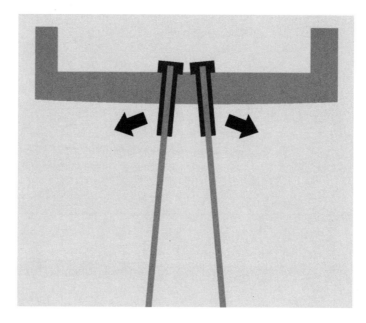

Figure 13-5 The rim is pulled in the direction of the spoke that is tightened.

Bent fork leg stays and dropouts are another common problem that you will find in many throwaway or scrap yard bicycle frames. Many times, bicycle frames found at the dump and scrap yards will have bends in the frame and forks. Traveling for miles under a ton of other scrap to then be dumped off the back of a truck is going to be a rough trip for a bicycle, and during this journey it will have suffered a few wounds. Since most of the frames you will find will be made of mild steel, bending the frames back into shape will not be a difficult task.

Figure 13-6 shows the three most common types of damage that you will find in unwanted frames—squashed stays and fork legs and bent rear dropouts. Most of these damages are minor and can be fixed using a little muscle to pull things back in place. Of course, if your frame is so severely bent that there are folds or creases in the tubes, then they will be un-repairable and only good for the scrap parts pile.

Straightening the rear dropouts is an easy task. Just place a large adjustable wrench onto the area that needs to be straightened, as shown in Figure 13-7, and force it back into place. If the dropouts have been bent so that the gap is too small for the rear axle to fit into, then pry them apart with a flathead screwdriver blade. For severe bends, you may need to place a block of wood under the area and bang out the dents with a hammer.

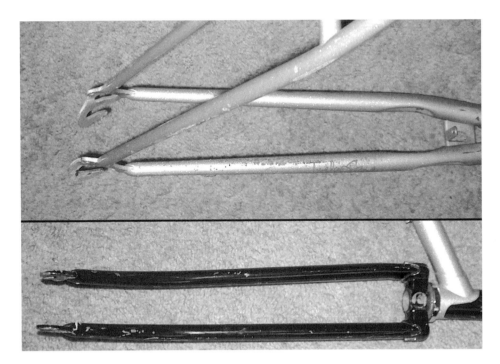

Figure 13-6   Many unwanted frames will have bent forks, stays, or dropouts.

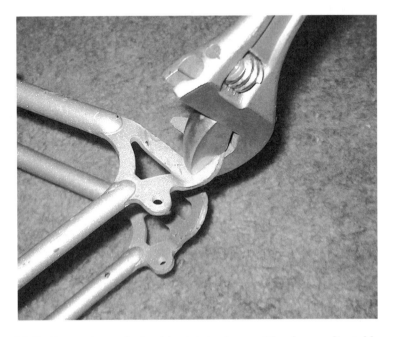

Figure 13-7   Dropouts can be bent back into shape with a large adjustable wrench.

This method of straightening will also work for crankset sprockets and fork dropouts as well. When attempting to straighten a sprocket, make sure it is mounted to the bike so you can give it a spin to check your work. If a crank arm has been bent, then don't bother trying to repair it, you will need an amazing amount of force to bend it back into place.

Rear stays that have been crushed together can be bent back into shape by using your own brute force. Place the side that is bent the most so it faces upward and place your foot on the bottom stay (see Figure 13-8). With both hands gripping the top stays, pull upward until they begin to bend back into shape. You will know you have put them back into alignment when you can place a rear wheel into the dropouts without having to squeeze or pull the stays in either direction to make it fit.

Bending the rear stays apart is not as hard as you might think, so take it easy and do it in steps so as to not bend them apart too far. If a frame is bent too many times, it will become weak and could fail. I do not recommend that you attempt to straighten an aluminum frame, as aluminum does not like to be bent and doing so will cause the bent area to become very brittle.

Bending the forks is done using the same method, but will require you to use a lot more strength, especially if trying to straighten a hefty set of BMX or mountain bike forks. The method used is the same

Figure 13-8 A little muscle power can be used to straighten the rear of a frame.

method used to straighten the rear stays—placing the worst leg upward while you hold down the other with your foot (see Figure 13-9). It's a good idea to do this operation in small steps, checking alignment often by looking down the length of the forks to make sure one leg is not bent too far. Use a front rim axle to gauge your progress, turning the forks around to pull on the other leg if necessary.

Checking the alignment often is important because you could widen the forks to the correct width and still end up with both legs out of alignment. Your goal is not just to make a front axle fit into the forks, but to make sure each leg is at an equal distance from the head tube. Heavier forks such as those found on BMX and mountain bikes are extremely strong, and may require a lot more force than you have to give. If you just can't seem to get the forks to bend, you could either have another person help you bend them, or place two long tubes over both fork legs and pry them apart by pulling the ends of the tubes apart. Remember to check your progress often, especially if using long tubes as leverage.

If your frame tubing is bent, then you are out of luck. Attempting to straighten a frame that has had a head-on collision is almost impossible, even if you heat the tube red hot with a torch. If you find a frame like this, strip off the parts and use the rest as scrap metal. Fixing a dent in a frame tube can be accomplished by either filling the area with auto body filler or by building up the area with beads of weld, then grinding it flush.

Figure 13-9   Pulling a set of front forks apart will require a lot of force.

Accessories and Eye Candy

Besides a nice paint job, nothing will make your bike stand out more than all the little trimmings and accessories. What would a chopper be without a phat rear fender or banana seat? What about your low-racer? You just have to put on a speedometer and a water bottle. Even a fresh set of tires and handle grips can make a big improvement in the final appearance of your project. So many builders create amazing new bicycle designs, then forget about adding the accessories, or even the paint!

Although the final project may be a great ride, it will fail to impress your friends if it looks rough, with unground welds or old parts thrown on. So avoid the urge to show off you machine or take pictures for your website before the final details have been added.

Tires, handle grips, cables, and chains are parts you will need to buy new, as old bikes usually do not have these parts in good condition. Handle grips usually need to be cut off the handlebars, cables are usually seized up solid, tires are usually worn out, and chains can be so rusted that no amount of soaking will bring them back to life.

Splurge a bit! If your final project actually turns out the way you want it to, then dress it up and show it off with cool accessories and a nice paint job.

BICYCLE-BUILDING RESOURCES

The Internet

Human-powered vehicle designers come from all walks of life and from all corners of the globe. It's always fun to see what other builders have done. Many times, just looking at someone else's design will spark your imagination and get your creative juices flowing. I once thought that strange bicycle designs such as tall bikes and ultralong forked choppers were pretty rare, but once I plugged my computer into the Internet, I realized there were thousands of other bike builders out there with ideas just as bizarre as my own.

In the year 2000, I launched my own Website, www.atomiczombie. com, to showcase some of my strange bicycle creations past and present and, within months, I was getting many e-mails from other builders from just about every known part of the world, and some places I never even heard of! It's always good to hear from other builders and share ideas and advice on what works and what doesn't.

There is a massive amount of knowledge on the Internet if you know how to find it.

The Internet will be your ultimate resource when it comes to finding information on every aspect of bicycle building and repair, but you have to know what to look for. Before I just spew out a long list of cool bike-related Websites, let me show you how to search the Internet to find what you want.

The first thing you need to do is log onto an Internet search engine. I recommend Google because it's the most powerful and fastest search engine on the Internet. In fact, most of the other search engines just reference Google, so you might as well go straight to the source. Type www.google.com into your browser's address bar and you will be presented with the interface shown in Figure 14-1.

Using this search engine is very straightforward. Simply type what you are looking for in the Address field, then click on Go or press Enter on the keyboard, and a pile of sites will be presented. The key to using a search engine is to refine your search to eliminate sites that have nothing to do with what you are looking for. If you type "atomic zombie" in the box and hit enter, my Website will be one of the first links that shows up. This is a fairly uncommon name, so the search results that are presented are fairly accurate.

If you try to do a search for something too general, you will have to look through a lot of links before you find what you want. Typing "chopper" in the search box will bring up hundreds of sites, some of which have nothing to do with two-wheelers at all. Chopper is a term that can describe many things—lawnmowers, custom motorcycles, electronic circuitry to convert DC to AC, someone's pet dog, etc, etc. If

Figure 14-1 Searching the Internet requires a good search engine like Google.

chopper bicycles is what you are looking for then try typing "chopper bicycles" in the box instead. Sounds reasonable, right?

If you narrow down your searches, you will find what you want a lot quicker without having to filter through unrelated and garbage links. There are also certain short forms that you will learn to help narrow your search to your field of interest. HPV is an acronym for Human Powered Vehicles and will bring up many sites with a host of good information. Of course, it will also bring up a few sites about "warts," but that's the Internet for you!

Some of my favorite links to bicycle builder's sites and clubs are listed below. At the time of writing, these sites were all working, but due to the ever-flowing river of information on the Internet, I cannot guarantee they will still be working when you try them, or if they will even be the same site. What was once "KCB—Karl's Cool Bikes" could now be "KCB—Kansas City Barbeque."

I tried to sort these links into some reasonable order, and this is only a small collection of all the links in my "favorites." If your link did not make it to the list, it doesn't mean I didn't enjoy your Website. I just missed it when I compiled this list.

Human-Powered Vehicle Sites and Clubs

Bike Rod and Custom Ezine
http://bikerodnkustom.homestead.com/

The Bicycle Forest
http://www.bikeforest.com/

Warren Beauchamp's Wisconsin HPV site
http://www.wisil.recumbents.com

Human-Powered Vehicles in Minnesota
http://mnhpva.org/

Human-Powered Vehicles in North Texas
http://www.rbent.org/

Human-Powered Vehicles in Detroit
http://www.lmb.org/wolbents/

Human-Powered Vehicles in New York
http://www.bluemoon.net/~padelbra/the_recumbenteers.htm

Human-Powered Vehicles in Southern Ontario
http://www.hpv.on.ca/

International Human-Powered Vehicle Association
http://www.ihpva.org/

Human-Powered Vehicles in Australia
http://sunsite.anu.edu.au/community/ozhpv/

Human-Powered Vehicles in Britain
http://www.bhpc.org.uk/

Human-Powered Vehicles in France
http://www.ihpva.org/chapters/france/

Human-Powered Vehicles in Denmark
http://www.dcf.dk/hpv/

Norwegian Human-Powered Vehicles
http://wwwserv2.iai.fzk.de/~wieland/nhpv.html

Human-Powered Vehicles in Japan
http://www.stmfr.co.jp/STMFR/club.html

Human-Powered Vehicles in New Zealand
http://www.converge.org.nz/hpvcanterbury/

Tasmanian HPV Enthusiast Club
http://sunsite.anu.edu.au/community/ozhpv/tas/index.htm

Queensland HPV Enthusiasts Group
http://sunsite.anu.edu.au/community/ozhpv/qldhpv/index.htm

A Human-Powered Airplane Called The Raven
http://ravenproject.org//

World Ice HPV Championships
http://www2.bitstream.net/~dkrafft/icebike/Ice99.html

Home Builder's Websites

Atomic Zombie Extreme Cycle Creations
http://www.atomiczombie.com

Troy Way's Home Page—The Blue Shark
http://www.bluepdoo.com

Mike Watson's Cool Creations
http://www.homestead.com/bikerodnkustom/watson.html

Sheldon Brown's Bicycle Information Site
http://sheldonbrown.org/bicycle.html

Barnett Williams' Recumbent Page
http://www.eng.uwaterloo.ca/~bkwillia/index.html

Tellef Øgrim from Norway
http://www.toegrim.freeservers.com/

Mark Bergstrom's Home-Built Bicycles
http://www.bergstrombicycles.com/

Marvin Penner's Bent Bikes
http://www.telusplanet.net/public/mcpenner/Marvinsbentbikes.html

The Cyclone by Robert den Hartigh
http://www.ee.ualberta.ca/~rob/cyclone.htm

Paul Walter's Home-Built Recumbents
http://www.elltel.net/recumbent/

Paul Keen's Home-Built Recumbents
http://www.users.bigpond.com/pekay/recum.htm

Chopper and Club Related Websites

Subversive Choppers Urban Legion
http://www.scul.org

Chunk Six-Six-Six Disinformation Distribution Cadre
http://www.dclxvi.org/chunk/

Human-Powered Creativity in Chicago
http://www.chicagofreakbike.org/

Skids Constructing Apocalyptic Bicycles
http://www.scab.freeservers.com/bikes.htm

Bike Hotrod—Cool Site with Many Choppers
http://www.bikehotrod.com/

Cyclecide Bike Club
http://www.cyclecide.com/

Revolution of Urban Self Transport
http://www.funkybikes.com/

Scallywags Bike Club
http://www.scallywagsbikeclub.com/

Custom Bicycle Designs
http://www.organicengines.com/

Choppercabras Bicycle Club
http://www.geocities.com/SunsetStrip/Bass/1520/

Ghetto-Fied Bikes
http://www.angelfire.com/punk3/gfbikes/

Chopper Riding Urban Dwellers
http://home.earthlink.net/~legume/

Dead Baby Bikes
http://www.geocities.com/Pipeline/1367/dbb.htm

KlunkerLeagueNow Klan
http://www.klunkerleaguenow.com/

Rat Patrol Bicycle Designs
http://www.geocities.com/ratpatrolhq/

Underlife BC, Hellsinki
http://www.helsinki.fi/~kytomaki/underlife/

This list makes up only a small fraction of the bicycle-related sites that are available on the Internet. If you have several days to spend locked in your room staring at your monitor, then visit the "links" section of a few of these sites and then visit the "links" section of those sites and you will find thousands of interesting bicycle sites. No matter what you can dream up, there is probably something even more bizarre on the Internet already!

Books of Interest

In addition to the Internet, books are probably your next-best source for bicycle-related information. Although a lot of the books you will find mainly deal with the maintenance and repair of the standard bicycle, they may still be of interest to you. After all, most of your projects use parts from a standard bicycle, so the information in these books will still apply. If you want to see a very long list of bicycle books, go to www.amazon.com and type in "bicycle" in the search box. The list is very long.

The following is a list of some of the bicycle related books that I have read and found interesting.

Encycleopedia
Alan Davidson and Jim McGurn
The Overlook Press; 5th edition
ISBN: 0879518847

Richard's 21st Century Bicycle Book
Richard Ballantine
The Overlook Press
ISBN: 1585671126

Bike Cult: The Ultimate Guide to Human-Powered Vehicles
David B. Perry
Four Walls Eight Windows
ISBN: 1568580274

The Bicycle Builder's Bible
Jack Wiley
Tab Books
ISBN: 0830611568

Building Your Perfect Bike: From Bare Frame to Personalized Superbike
Richard Ries
Van der Plas Publications
ISBN: 093320177X

The Fantastic Bicycles Book
Steven Lindblom
Houghton Mifflin Co. (Juvenile)
ASIN: 0395284813

Extreme Bicycle Stunt Riding Moves (Behind the Moves)
Danny Parr
Capstone Press
ISBN: 0736807810

The Bicycle Wheel
Jobst Brandt
Avocet; 3rd edition
ISBN: 0960723668

Bicycle Design
Mike Burrows
Alpenbooks
ISBN: 0966979524

I hope you find all the information you are looking for! Don't forget the best source of information of all—hands-on experience. If you can't find something on a particular design or idea, then get out to the workshop and try it for yourself!

INDEX

About the Authors

Brad Graham is a 34-year-old, self-employed computer network engineer and dad from Thunder Bay, Ontario, Canada. So, what's a computer nerd doing writing a book on radical bicycle design and customization, you ask? Since the age of 10, Brad has never been satisfied with owning a "normal," two-wheeled bicycle. He has joined more and more junk together in an effort to rise above conformity and be different. Although friends and neighbors agree that Brad's strange inventions have always been a source of entertainment, in his mind he always knew that all of this hard work would someday turn into a lifetime hobby.

With no more pieces held together with duct tape or bolts, Brad's crazy bicycle creations began to take on more radical forms. Bikes grew taller, longer, faster, and they lasted longer than a few hours on the street. Almost anything that could be drawn on a piece of paper could be hacked into a working prototype if enough parts were welded to it. Tandems, tall bikes, trikes, unicycles, recumbents—all of these bikes were coming out of the garage just as fast as the junk went in. Brad's workshop was like a bicycle factory. Old rusty bikes would go in on one conveyer belt, and strange, amazing machines would come out on another conveyer belt, just like the wacky bikes in Dr. Seuss's books and cartoons.

Not all of the bikes Brad created were simply random expressions of total chaos. Some were actually quite functional, and lasted quite a long time before returning to the giant pile of junk in the basement. Take, for instance, the handmade tandem bicycle made from two old road bikes. This bike was used for almost two years. It had six speeds, a nice red paint job, and brakes, which was a rare feature for many of his projects in those days.

After two years, and bored with the tandem design, Brad set off to make a real HPV (human-powered vehicle). An HPV is a decent replacement for a car for those who seek the benefits of exercise or want to help cut down on emissions. But, for a 15-year-old, an HPV was the only option, since the prospect of owning a real car was still a year away. This trike had many of the features of a car, and an original car seat from a 1973 Chevy Nova (Figure 1). He could lie back in the comfortable seat with one hand on the remote steering wheel and cruise all around town. Due to the bike's recumbent position, he could reach fairly high speeds as well, which was another feature Brad often pursued.

Considered by some to be the wackiest of Brad's bicycle creations, the Skycycle is a machine that is both awesome and terrifying. At 8 feet tall, this slender, ladder-like bike was ridden on only two wheels and without brakes by only the hardcore adrenaline-thrill-seekers, which, of course, included Brad and most of his friends (Figure 2).

Figure 1   Brad's first Human-Powered Vehicle (HPV), with a 1973 Chevy Nova car seat.

Figure 2   Brad's first Skycycle was created in the mid-1980s. Courtesy of *The Chronicle-Journal*.

Unfortunately, like many teens, the introduction of the car makes it easy to forget the trusty old bike, and soon the only thing that mattered was horsepower and "mag" thickness. Many years passed before Brad began to remember the fun that could be had on a bike, and it wasn't until the age of 32 that, once again, the passion to go as fast as possible without a motor called on him.

In the meantime, he became an accomplished welder, and found the ultimate source for free bike parts—the big metal scrap pile at the city landfill site—so the rest was easy. Thanks to the expansion of the Internet, finding other crazy bike builders was easy, and the amount of new ideas that could be found for inspiration was amazing.

Brad built a few new designs, took a lot of pictures along the way, and posted them on his own Website at www.atomiczombie.com just for fun. Within a few months, the Website was getting flooded with hits, and his e-mail box was constantly jammed with e-mails from other custom bike builders or new enthusiasts who were anxious to get started. A year and many projects later, the Website is known internationally, and the idea for this book was born.

Kathy McGowan was born and raised in Thunder Bay, Ontario, Canada. She has always had an interest in the publishing industry. After obtaining her BA in psychology and a journalism diploma in Ottawa, Ontario, Kathy returned to her hometown in the mid-1990s to pursue various media and communications opportunities. In the spring of 2000, she reconnected with childhood friend, Brad Graham, and later that year the pair formed their own computer sales, service, networking, and training company.

Last fall, Kathy and Brad embarked on another venture together—to write a user-friendly, step-by-step book containing hundreds of pictures to help other bike enthusiasts build their own creations. In addition to posing for pictures, editing chapters, and general project management for *Atomic Zombie's Bicycle Builder's Bonanza,* Kathy manages their company's daily operations. She is looking forward to working with Brad, her partner in business and in life, on another book in the near future. She loves cruising the streets on her custom Kool Kat along with the Blue Shark and Marauder guys!